Content

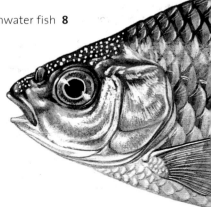

Contents (cont.)

hamlyn

r

d by Keith Linsell

Preface

The main aim of this book is to provide a short popular account of the freshwater fish of Europe, together with a means of identifying them. Because of the large number of species involved only the major features of the biology of each can be presented, and this is done in an entirely consistent manner among species in order to facilitate comparisons. It is hoped that this guide will stimulate interest in Europe's attractive and varied fish fauna not only among those directly concerned, such as anglers, aquarists, conservationists and fishery biologists, but also among those indirectly involved, such as agriculturalists, industrialists and politicians, whose thoughts and decisions are among the main factors affecting the future nature of Europe's fish populations.

In creating this guide I have been greatly assisted and encouraged by colleagues and correspondents in many different countries. Special thanks must go to Yoichi Machino for his advice and substantial help in obtaining literature. I am grateful also for help from Suleyman Balik, Patrick Berrebi, Alain Crivelli, Panos Economidis and Peter Miller. The magnificent colour illustrations by Keith Linsell represent, I believe, the finest set of paintings of European freshwater fish ever executed, and are an enormous asset to a guide of this kind. The line drawings by Jennifer Middleton maintain the same high standard. I would like to thank other artists for their care and interest in the work. The help I have received throughout the development of the book from of Hamlyn has greatly facilitated its production and I am grateful for their patience and advice.

Peter S. Maitland
Haddington, Scotland, 2000

Executive Editor: Julian Brown
Project Editor: Tarda Davison-Aitkins
Editor: Adam Ward
Creative Director: Keith Martin
Picture Research: Wendy Gay
Production: Sarah Scanlon

First published in Great Britain in 1977
Revised and updated in 2000
by Hamlyn, an imprint of
Octopus Publishing Group Limited
2–4 Heron Quays
London E14 4JP

Copyright © 2000 Octopus Publishing Group Limited

ISBN 0 600 59690 7

A catalogue record for this book is available from the British Library

Design @ 2wo
Produced by Grafos S.A
Printed in Spain

How to use this book

This guide covers the continent of Europe bordered by the Arctic and Atlantic Oceans, the Mediterranean and Black Seas, the Caucasus Mountains, the Caspian Sea, and the Ural River and Mountains. (see fig. 1). This whole area covers a land mass of more than 10 million km^2 (3,900,000 ml^2) and includes a very wide range of freshwater habitats, from cold deep northern lakes and rivers, some of which are frozen for much of the year, to warm shallow ponds and streams in the Mediterranean area which are never frozen.

The book is divided into four broad sections. First there is a general account of fish – their structure and physiology, behaviour and development, and ecology and distribution. This is concluded by a study of their value to man, not only commercially, but also from sporting, aesthetic, and scientific viewpoints. The second section is an identification key to the families of fish found in fresh water in Europe. This is based on the standard dichotomous pattern found in many keys, where at each point in the key the reader who is trying to identify a fish is presented with two alternatives. The one followed is that which fits the specimen best. Having selected one alternative the reader is then led to another couplet, and this continues until the identity of the family is reached. This can generally then be confirmed by reference to the outline diagram provided for each family. The third section of the book is probably the most important, and is a guide to the individual species, family by family. Having identified the family from the first key, the reader then goes to a similar type of key within each family section to make the final identification. This can then be confirmed by the colour illustration and to some extent by the distribution map – where for each species inland distribution is shown in pink, and coastal distribution is shown in blue. The final section of the book is a list of useful references for further reading – usually these relate to particular aspects of fish biology or to accounts of the fish fauna of one particular country – an illustrated glossary, and an index (see pages 250–256).

Formal identification using the whole key is often not necessary, especially with the more unusual or obvious species. Thus if the reader wishes to identify an eel-like fish with one pair of fins it can be seen immediately from pages 56–61, that this must be in the family Anguillidae. On looking up the relevant page in the species accounts it will be found that if it is from fresh water in Europe it can really only be one species – the common European Eel (*Anguilla anguilla*).

The common names in this guide are those in general use, and each is followed by the appropriate scientific name. This consists of two parts: the genus (equivalent to a surname in Europe), and the specific name (equivalent to a christian name). It should be noted, however, that the specific name is always placed second. Where no common names are available for a species, the most acceptable foreign name, or its translation is given. For each species, the text, distribution map and illustration are placed together. The text has been standardised as much as possible to facilitate comparisons, and includes information on the size, habitat, distribution, breeding habits, growth, food, value and conservation status of each species. The maps show the freshwater distribution, and where applicable, the coastal distribution. It must be remembered that each species occurs only in suitable types of water within the areas shown, and not everywhere. There is normally only one illustration for each species, except where there are major differences between the sexes, for instance in the Three-spined Stickleback, or specific colour varieties, for instance in the Orfe. Where both sexes are illustrated, ♂ denotes male, and ♀ denotes female.

Key to distribution map: ▓ inland distribution ▓ coastal distribution

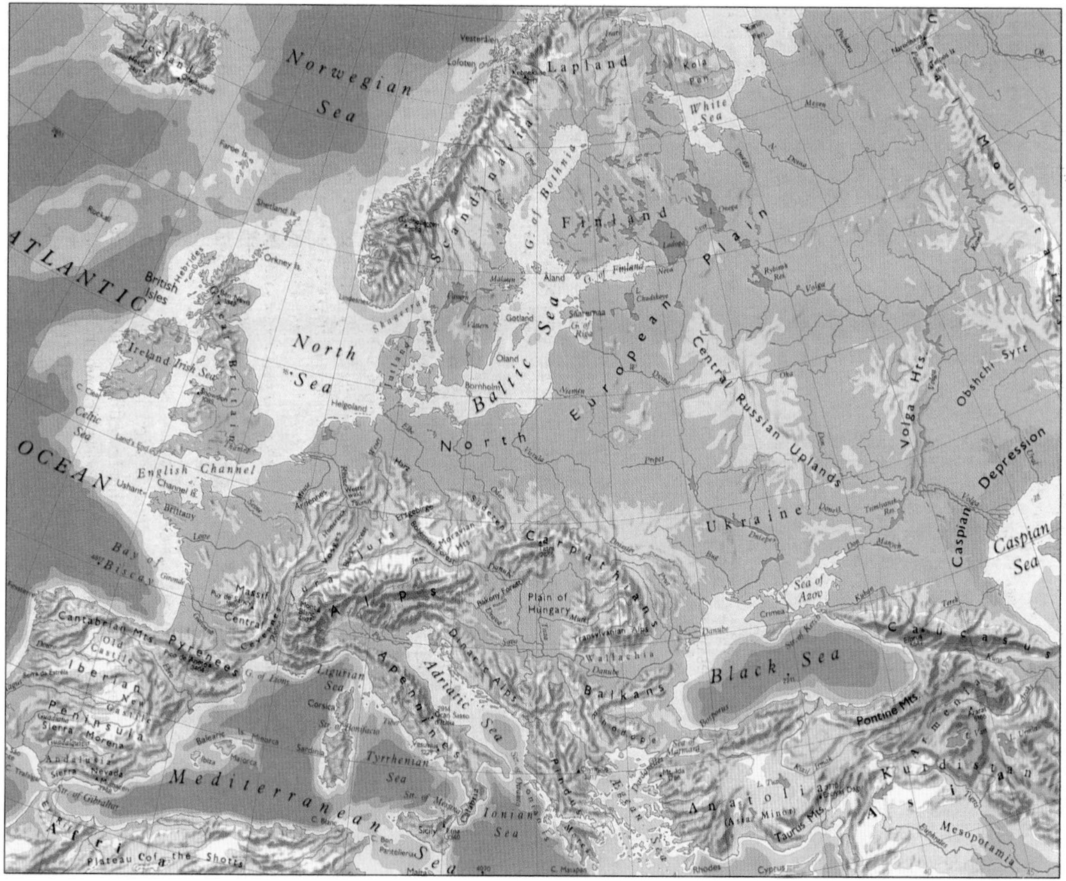

Fig. 1 Map of Europe showing some important rivers, lakes and mountain ranges.

© George Philip Limited

6

Introduction

The classification scheme used for fish families in this book follows that of Nelson (1994). The scientific names follow the review by Kottelat (1997), except for the Salmonidae, Coregonidae and a few other families where more conservative nomenclature is adopted. There are many other books written on various aspects of the freshwater fish of Europe, and the more important of these are included in the bibliography at the end of this guide. Each of these may in turn contain useful references to specific topics. Identification of many species found in Europe is often possible using certain of these publications, simply by reference to, and comparison with, the various illustrations in them. This is often a slow and inaccurate method of identification and, moreover, most of the older works on the freshwater fish of Europe do not include those species introduced relatively recently, for instance Largemouth Bass (*Micropterus salmoides*) and Pink Salmon (*Oncorhynchus gorbuscha*) from North America. It is hoped that this guide will prove to be a relatively simple but accurate means of identifying any species of fish found regularly in fresh or brackish water in Europe.

This guide includes animals belonging to both cyclostomes and Pisces. Technically, since the cyclostomes belong to the Agnatha and have therefore no proper jaws, only the latter are true fish. However, in this book the term 'fish' is understood to include both cyclostomes and Pisces.

Unfortunately we are still very ignorant about many European fish, and indeed it is doubtful if a number of the species included here are proper species. It is more likely that they are simply local races of a more widespread species. However, until the groups concerned have been reviewed by competent taxonomists, most of these 'species' are dealt with individually.

In addition to those species of fish which are known to occur in fresh waters in Europe, a few other categories are also included in this guide. The most important among these are brackish water forms which, although basically marine, are also found regularly in brackish and sometimes in fresh waters. Marine species which occur only infrequently in brackish water are generally not included in the key, although mention is made of a few of these fish where there may be confusion with true brackish water forms. Of necessity, the decision whether or not to include some of these species has been rather arbitrary. Native forms make up the bulk of the European freshwater fish fauna, but there are also several introduced species which have thrived and bred in this continent, sometimes establishing populations over quite a wide area. All such species are included in the keys that follow. Other species of freshwater fish of doubtful status in Europe have not been included: among these are various tropical species associated with heated effluents, like the Guppy (*Poecilia reticulata*) and odd records of temperate species for which there is no evidence of the existence of a population. These are discussed further below.

This guide is intended for use in the field, so it should be feasible to return fish to the water alive after capture, examination and identification. During this process specimens should always be kept as cool and as damp as possible. The features used to differentiate between families and between species are mainly external; and in this book characters which are as objective and absolute as possible have been selected where feasible. Nevertheless, it is necessary in some instances to resort to features which involve killing and dissecting the specimen; the characters concerned here are mainly found in the region of the head – pharyngeal bones (in Cyprinidae), vomer bones (in Salmonidae) and gills (in Clupeidae), etc.

As already mentioned, the keys given follow the dichotomous pattern common to many field guides. Where possible, several distinguishing characters have been used at each point in the key and these should be considered in combination with each other. Most species are illustrated, and both the text and the relevant illustration should be consulted. Due consideration must always be given to the possibility of any specimen being very young, malformed in some way, or a hybrid, vagrant or introduced species. Doubtful or little known species are not dealt with in full and indicated *.

The most common numerical features used in the key are counts of fin rays and of scales. In the fin ray counts, the number for each fin may include both the branched and the unbranched rays. The main scale counts are taken along the lateral line, starting at the first scale behind the operculum and ending at the last scale before the caudal fin. Some diagonal scale counts may also be used: these are normally counted from the lateral line up to the adipose fin (where present), and from the lateral line down to the anal fin. Occasionally, counts are made from the lateral line up to the dorsal fin, and down to the pelvic fin.

Where colours are used in the key they refer to the condition in the fresh fish and should be true irrespective of size (above the larva and fry stages), sex and condition, unless otherwise stated. With many species it is possible to determine the sex accurately only by dissection of the sex organs (this is especially true outside the breeding season); with others there are constant external sexual differences.

Various species of fish in Europe hybridise quite frequently in the wild. Such hybrids are difficult to identify with a key such as is found here, for normally their characters are intermediate between those of the two parent species. Unfortunately, due to the very nature of speciation it is often those species of fish which are most alike (and thus difficult to separate) which are most likely to hybridise. A very wide variety of hybrids has been recorded from Europe, mostly in the family Cyprinidae.

Check list of European freshwater fish

Lamprey Family Petromyzontidae
Caspian Lamprey *Caspiomyzon wagneri* (Kessler 1870)
Danube Lamprey *Eudontomyzon danfordi* Regan 1911
Greek Lamprey *Eudontomyzon hellenicus* Vladykov, Renaud, Kott & Economidis 1982
Ukrainian Lamprey *Eudontomyzon mariae* (Berg 1931)
Vladykov's Lamprey *Eudontomyzon vladykovi* Oliva & Zanandrea 1959
River Lamprey *Lampetra fluviatilis* (Linnaeus 1758)
Brook Lamprey *Lampetra planeri* (Bloch 1784)
Arctic Lamprey *Lethenteron camtschaticum* (Tilesius 1811)
Lombardy Lamprey *Lethenteron zanandreai* (Vladykov 1955)
Sea Lamprey *Petromyzon marinus* Linnaeus 1758

Sturgeon Family Acipenseridae
Siberian Sturgeon *Acipenser baeri* Brandt 1869
Russian Sturgeon *Acipenser gueldenstaedtii* Brandt & Ratzeburg 1833
Adriatic Sturgeon *Acipenser naccarii* Bonaparte 1836
Ship Sturgeon *Acipenser nudiventris* Lovetsky 1828
Sterlet *Acipenser ruthenus* Linnaeus 1758
Stellate Sturgeon *Acipenser stellatus* Pallas 1771
Atlantic Sturgeon *Acipenser sturio* Linnaeus 1758
Beluga Sturgeon *Huso huso* (Linnaeus 1758)

Eel Family Anguillidae
European Eel *Anguilla anguilla* (Linnaeus 1758)

Herring Family Clupeidae
Allis Shad *Alosa alosa* (Linnaeus 1758)
Twaite Shad *Alosa fallax* (La Cepede 1803)
Macedonian Shad *Alosa macedonica* (Vinciguerra 1921)
Caspian Shad *Caspialosa caspia* Eichwald 1838
Kizilagach Shad *Caspialosa curensis* Surowow 1904
Dolginka Shad *Caspialosa maeotica* (Grimm 1901)
Black Sea Shad *Caspialosa pontica* Eichwald 1838
Bigeye Shad *Caspialosa saposhnikovi* Grimm 1901
Suworow's Shad *Caspialosa suworowi* (Berg 1913)
Abrau Sardelle *Clupeonella abrau* (Maliatski 1931)
Tyulka Sardella *Clupeonella cultriventris* (Nordmann 1840)

Carp Family Cyprinidae

Blue Bream *Abramis ballerus* (Linnaeus 1758)
Silver Bream *Abramis bjoerkna* (Linnaeus 1758)
Common Bream *Abramis brama* (Linnaeus 1758)
Whiteye Bream *Abramis sapa* (Pallas 1814)
Schneider *Alburnoides bipunctatus* (Bloch 1782)
White Bleak *Alburnus albidus* (Costa 1838)
Common Bleak *Alburnus alburnus* (Linnaeus 1758)
Caucasian Bleak *Alburnus charusini* Herzenstein 1889
Iberian Minnow *Anaecypris hispanica* (Steindachner 1866)
Asp *Aspius aspius* (Linnaeus 1758)
Dalmatian Barbelgudgeon *Aulopyge hugeli* Heckel 1842
Albanian Barbel *Barbus albanicus* Steindachner 1870
Common Barbel *Barbus barbus* (Linnaeus 1758)
Boca Barbel *Barbus bocagei* Steindachner 1865
Aral Barbel *Barbus brachycephalus* Berg 1914
Dog Barbel *Barbus caninus* Bonaparte 1839*
Bulatmai Barbel *Barbus capito* Gueldenstaedt 1773
Caucasian Barbel *Barbus ciscaucasicus* Kessler 1877
Iberian Barbel *Barbus comizo* Steindachner 1865
Turkish Barbel *Barbus cyclolepis* Heckel 1837
Euboean Barbel *Barbus euboicus* Stephanidis 1950*
Greek Barbel *Barbus graecus* Steindachner 1896
Graells' Barbel *Barbus graellsii* Steindachner 1866*
Guiraonis Barbel *Barbus guiraonis* Steindachner 1866*
Orange Barbel *Barbus haasi* Mertens 1924*
Macedonian Barbel *Barbus macedonicus* Karaman 1928*
Mediterranean Barbel *Barbus meridionalis* Risso 1826
Smallhead Barbel *Barbus microcephalus* Almaca 1967*
Peloponnesian Barbel *Barbus peloponnesius* Valenciennes 1842*
Adriatic Barbel *Barbus plebejus* Bonaparte 1839*
Prespa Barbel *Barbus prespensis* Karaman 1924
Sclater Barbel *Barbus sclateri* Gunther 1868*
Steindachner's Barbel *Barbus steindachneri* Almaca 1967*
Tyber Barbel *Barbus tyberinus* Bonaparte 1839*
Goldfish *Carassius auratus* (Linnaeus 1758)
Crucian Carp *Carassius carassius* (Linnaeus 1758)
Prespa Bleak *Chalcalburnus belvica* (Karaman 1924)
Shemaya Bleak *Chalcalburnus chalcoides* (Gueldenstaedt 1772)
Arrigon Nase *Chondrostoma arrigonis* (Steindachner 1866)*
Caucasian Nase *Chondrostoma colchicum* Kessler 1899
Durien Nase *Chondrostoma duriense* Coelho 1985*
Laska Nase *Chondrostoma genei* (Bonaparte 1839)
Dalmatian Nase *Chondrostoma knerii* Heckel 1843
Kuban Nase *Chondrostoma kubanicum* Berg 1914*
Spanish Nase *Chondrostoma miegii* Steindachner 1866*
Common Nase *Chondrostoma nasus* (Linnaeus 1758)
Terek Nase *Chondrostoma oxyrhynchum* Kessler 1877
Minnow Nase *Chondrostoma phoxinus* Heckel 1843
Iberian Nase *Chondrostoma polylepis* Steindachner 1865
Prespa Nase *Chondrostoma prespense* Karaman 1924*
Skadar Nase *Chondrostoma scodrense* Elvira 1987*
Italian Nase *Chondrostoma soetta* Bonaparte 1840
French Nase *Chondrostoma toxostoma* (Vallot 1837)
Turia Nase *Chondrostoma turiense* Elvira 1987*
Vardar Nase *Chondrostoma vardarense* Karaman 1928*
Variable Nase *Chondrostoma variabile* Jakowlew 1870*
Willkomm's Nase *Chondrostoma willkommii* Steindachner 1866*
Grass Carp *Ctenopharyngodon idella* (Valenciennes 1844)
Common Carp *Cyprinus carpio* Linnaeus 1758
Swamp Minnow *Eupallasella perenurus* (Pallas 1841)
Whitefin Gudgeon *Gobio albipinnatus* Lukasch 1933
Banarescu's Gudgeon *Gobio banarescui* Dimovski & Grupce 1974*
Italian Gudgeon *Gobio benacensis* (Pollini 1816)*
Caucasian Gudgeon *Gobio ciscaucasicus* Berg 1932
Aliakmon Gudgeon *Gobio elimeius* Kattoulas, Stephanidis & Economidis 1973*
Common Gudgeon *Gobio gobio* (Linnaeus 1758)

Kessler's Gudgeon *Gobio kessleri* Dybowski 1862
Danube Gudgeon *Gobio uranoscopus* (Agassiz 1828)
Silver Carp *Hypophthalmichthys molitrix* (Valenciennes 1844)
Bighead Carp *Hypophthalmichthys nobilis* (Richardson 1845)
Iberian Carpdace *Iberocypris palaciosi* Diadro 1980
Rhodos Carplette *Ladigesocypris ghigii* (Gianferrari 1927)*
Belica *Leucaspius delineatus* (Heckel 1843)
Caucasian Chub *Leuciscus aphipsi* Aleksandrov 1927
Black Sea Chub *Leuciscus borysthenicus* (Kessler 1859)
Burdig Dace *Leuciscus burdigalensis* Valenciennes 1844*
Carol Dace *Leuciscus carolitertii* Doadrio 1988*
Common Chub *Leuciscus cephalus* (Linnaeus 1758)
Danilewskii's Dace *Leuciscus danilewskii* (Kessler 1877)
Orfe *Leuciscus idus* (Linnaeus 1758)
Yugoslavian Dace *Leuciscus illyricus* (Heckel & Kner 1858)
Sparta Dace *Leuciscus keadicus* Stephanidis 1971*
Common Dace *Leuciscus leuciscus* (Linnaeus 1758)
Italian Dace *Leuciscus lucumonis* Bianco 1983*
Makal Dace *Leuciscus microlepis* (Heckel 1843)
Montenegran Dace *Leuciscus montenigrinus* Vidovic 1963*
Piedmont Dace *Leuciscus muticellus* Bonaparte 1837*
Croatian Dace *Leuciscus polylepis* (Steindachner 1866)
Pyrenean Dace *Leuciscus pyrenaicus* Gunther 1868
Blageon *Leuciscus souffia* Risso 1826
Adriatic Dace *Leuciscus svallize* (Heckel & Kner 1858)
Turskyi Dace *Leuciscus turskyi* (Heckel 1843)
Ukliva Dace *Leuciscus ukliva* (Heckel 1843)
Zrmanjan Dace *Leuciscus zrmanjae* (Karaman 1928)*
Macedonian Moranec *Pachychilon macedonicum* (Steindachner 1892)*
Moranec *Pachychilon pictum* (Heckel & Kner 1858)
Chekhon *Pelecus cultratus* (Linnaeus 1758)
Spotted Minnow *Phoxinellus adspersus* (Heckel 1843)
Adriatic Minnow *Phoxinellus alepidotus* Heckel 1843
Croatian Minnow *Phoxinellus croaticus* Steindachner 1866
Greek Minnow *Phoxinellus epiroticus* (Steindachner 1896)
Cave Minnow *Phoxinellus fontinalis* Karaman 1972*
Dalmatian Minnow *Phoxinellus ghetaldii* (Steindachner 1882)
Karst Minnow *Phoxinellus metohiensis* (Steindachner 1901)*
Speckled Minnow *Phoxinellus pleurobipunctatus* (Stephanidis 1939)*
Prespa Minnow *Phoxinellus prespensis* (Karaman 1924)*
South Dalmatian Minnow *Phoxinellus pstrossi* (Steindachner 1882)
Poznan Minnow *Phoxinus czekanowskii* Berg 1932
Common Minnow *Phoxinus phoxinus* (Linnaeus 1758)
Fathead Minnow *Pimephales promelas* Rafinesque 1820
Yliki Minnowcarp *Pseudophoxinus beoticus* (Stephanidis 1939)*
Albanian Minnowcarp *Pseudophoxinus minutus* (Karaman 1824)
Marida Minnowcarp *Pseudophoxinus stymphalicus* (Valenciennes 1844)
False Harlequin *Pseudorasbora parva* (Temminck & Schlegel 1842)
Bitterling *Rhodeus sericeus* (Bloch 1782)
Calandino Roach *Rutilus alburnoides* (Steindachner 1866)
Galician Roach *Rutilus arcasii* (Steindachner 1866)*
Alpine Roach *Rutilus aula* (Bonaparte 1841)*
Dalmatian Roach *Rutilus basak* (Heckel 1843)*
Pearl Roach *Rutilus frisii* (Nordmann 1840)
Heckel's Roach *Rutilus heckelii* (Nordmann 1840)*
Karaman's Roach *Rutilus karamani* Fowler 1977*
Pardilla Roach *Rutilus lemmingii* (Steindachner 1866)
Macedonian Roach *Rutilus macedonicus* (Steindachner 1892)
Portuguese Roach *Rutilus macrolepidotus* (Steindachner 1866)
Austrian Roach *Rutilus meidingeri* (Heckel 1851)*
Ohrid Roach *Rutilus ohridanus* (Karaman 1924)*
Danube Roach *Rutilus pigus* (La Cepede 1803)
Prespa Roach *Rutilus prespensis* (Karaman 1924)*
Adriatic Roach *Rutilus rubilio* (Bonaparte 1837)
Common Roach *Rutilus rutilus* (Linnaeus 1758)
Yliki Roach *Rutilus ylikiensis* Economidis 1991*

Acheloos Rudd *Scardinius acarnanicus* Economidis 1991*
Common Rudd *Scardinius erythrophthalmus* (Linnaeus 1758)
Greek Rudd *Scardinius graecus* Stephanidis 1937
Rumanian Rudd *Scardinius racovitzai* Muller 1958*
Adriatic Rudd *Scardinius scardafa* (Bonaparte 1837)*
Tench *Tinca tinca* (Linnaeus 1758)
Hellenic Minnowroach *Tropidophoxinellus hellenicus* (Stephanidis 1939)
Spartian Minnowroach *Tropidophoxinellus spartiaticus* (Schmidt-Ries 1943)
River Vimba *Vimba elongata* (Valenciennes 1844)*
Dark Vimba *Vimba melanops* (Heckel 1837)
Common Vimba *Vimba vimba* (Linnaeus 1758)

Spined Loach Family Cobitidae
Arachthos Loach *Cobitis arachthosensis* Economidis & Nalbant 1997*
Twostriped Loach *Cobitis bilineata* (Canestrini 1866)*
Lamprehuela Loach *Cobitis calderoni* Bacescu 1962*
Caspian Loach *Cobitis caspia* Eichwald 1838
Caucasian Loach *Cobitis caucasica* Berg 1906
Venetian Loach *Cobitis conspersa* (Cantoni 1882)
Balkan Loach *Cobitis elongata* Heckel & Kner 1858
Slender Loach *Cobitis elongatoides* Bacescu & Maier 1969*
Hellenic Loach *Cobitis hellenica* Economidis & Nalbant 1997*
Moroccan Loach *Cobitis maroccana* (Pellegrin 1929)
Delta Loach *Cobitis megaspila* Nalbant 1993*
Prespa Loach *Cobitis meridionalis* Karaman 1924
Marsh Loach *Cobitis paludica* (de Buen 1930)*
Striped Loach *Cobitis punctilineata* Economidis & Nalbant 1997*
Stephanidis' Loach *Cobitis stephanidisi* Economidis 1992*
Strumica Loach *Cobitis strumicae* Karaman 1955*
Spined Loach *Cobitis taenia* Linnaeus 1758
Trichonis Loach *Cobitis trichonica* Stephanidis 1974*
Vardar Loach *Cobitis vardarensis* Karaman 1928*
Weather Loach *Misgurnus fossilis* (Linnaeus 1758)
Golden Loach *Sabanejewia balcanica* (Karaman 1922)
Bulgarian Loach *Sabanejewia bulgarica* (Drensky 1928)*
Italian Loach *Sabanejewia larvata* (Filippi 1859)
Rumanian Loach *Sabanejewia romanica* (Bacesco 1943)

Stone Loach Family Balitoridae
Angora Loach *Barbatula angorae* (Steindachner 1897)
Stone Loach *Barbatula barbatula* (Linnaeus 1758)
Struma Loach *Barbatula bureschi* (Drensky 1928)*
Terek Loach *Barbatula merga* (Krynicki 1840)

American Catfish Family Ictaluridae
Black Bullhead *Ameiurus melas* (Rafinesque 1820)
Brown Bullhead *Ameiurus nebulosus* (Lesueur 1819)
Channel Catfish *Ictalurus punctatus* (Rafinesque 1818)

European Catfish Family Siluridae
Aristotle's Catfish *Silurus aristotelis* Garman 1890
Wels Catfish *Silurus glanis* Linnaeus 1758

Pike Family Esocidae
Northern Pike *Esox lucius* Linnaeus 1758

Mudminnow Family Umbridae
European Mudminnow *Umbra krameri* Walbaum 1792
Eastern Mudminnow *Umbra pygmaea* (De Kay 1842)

Smelt Family Osmeridae
European Smelt *Osmerus eperlanus* (Linnaeus 1758)

Whitefish Family Coregonidae
Vendace *Coregonus albula* (Linnaeus 1758)
Arctic Cisco *Coregonus autumnalis* (Pallas 1811)
Common Whitefish *Coregonus lavaretus* (Linnaeus 1758)
Broad Whitefish *Coregonus nasus* (Pallas 1776)
Houting *Coregonus oxyrinchus* (Linnaeus 1758)
Peled *Coregonus peled* (Gmelin 1788)

Arctic Whitefish *Coregonus pidschian* (Gmelin 1788)
Sheefish *Stenodus leucichthys* (Gueldenstaedt 1772)

Salmon Family Salmonidae
Huchen *Hucho hucho* (Linnaeus 1758)
Pink Salmon *Oncorhynchus gorbuscha* (Walbaum 1792)
Chum Salmon *Oncorhynchus keta* (Walbaum 1792)
Coho Salmon *Oncorhynchus kisutch* (Walbaum 1792)
Rainbow Trout *Oncorhynchus mykiss* (Walbaum 1792)
Marbled Trout *Salmo marmoratus* Cuvier 1829
Atlantic Salmon *Salmo salar* Linnaeus 1758
Brown Trout *Salmo trutta* Linnaeus 1758
Adriatic Salmon *Salmothymus obtusirostris* (Heckel 1852)
Arctic Charr *Salvelinus alpinus* (Linnaeus 1758)
American Brook Charr *Salvelinus fontinalis* (Mitchill 1814)
American Lake Charr *Salvelinus namaycush* (Walbaum 1792)

Grayling Family Thymallidae
European Grayling *Thymallus thymallus* (Linnaeus 1758)

Cod Family Gadidae
Burbot *Lota lota* (Linnaeus 1758)

Grey Mullet Family Mugilidae
Thicklipped Grey Mullet *Chelon labrosus* Risso 1826
Golden Grey Mullet *Liza aurata* (Risso 1810)
Thinlipped Grey Mullet *Liza ramada* (Risso 1826)
Sharpnose Grey Mullet *Liza saliens* (Risso 1810)
Striped Grey Mullet *Mugil cephalus* Linnaeus 1758
Boxlipped Grey Mullet *Oedalechilus labeo* Cuvier 1828

Silverside Family Atherinidae
Bigscale Silverside *Atherina boyeri* Risso 1810
Common Silverside *Atherina hepsetus* Linnaeus 1758
Mediterranean Silverside *Atherina mochon* Valenciennes 1835
Atlantic Silverside *Atherina presbyter* Cuvier 1829
Pejerrey Silverside *Odonthestes bonariensis* (Valenciennes 1835)

Killifish Family Fundulidae
Mummichog *Fundulus heteroclitus* (Linnaeus 1766)

Valencia Family Valenciidae
Spanish Valencia *Valencia hispanica* (Valenciennes 1846)
Greek Valencia *Valencia letourneuxi* (Sauvage 1880)

Toothcarp Family Cyprinodontidae
Fasciated Toothcarp *Aphanius fasciatus* (Valenciennes 1821)
Iberian Toothcarp *Aphanius iberus* (Valenciennes 1846)

Livebearer Family Poeciliidae
Mosquito Fish *Gambusia affinis* Baird & Girard 1853

Stickleback Family Gasterosteidae
Three-spined Stickleback *Gasterosteus aculeatus* Linnaeus 1758
Greek Stickleback *Pungitius hellenicus* Stephanidis 1971
Ukrainian Stickleback *Pungitius platygaster* (Kessler 1859)
Nine-spined Stickleback *Pungitius pungitius* (Linnaeus 1758)

Pipefish Family Syngnathidae
Straightnose Pipefish *Nerophis ophidion* (Linnaeus 1758)
Shore Pipefish *Syngnathus abaster* Risso 1826
Greater Pipefish *Syngnathus acus* Linnaeus 1758
Deepnose Pipefish *Syphonostoma typhle* Linnaeus 1758

Sculpin Family Cottidae
Common Bullhead *Cottus gobio* Linnaeus 1758
Siberian Sculpin *Cottus poecilopus* Heckel 1837
Fourhorn Sculpin *Triglopsis quadricornis* (Linnaeus 1758)

Bass Family Moronidae
Sea Bass *Dicentrarchus labrax* (Linnaeus 1758)
Spotted Bass *Dicentrarchus punctatus* Bloch 1797

Sunfish Family Centrarchidae
Rock Bass *Ambloplites rupestris* Rafinesque 1817
Redbreast Sunfish *Lepomis auritis* (Linnaeus 1758)
Green Sunfish *Lepomis cyanellus* (Rafinesque 1819)
Pumpkinseed *Lepomis gibbosus* (Linnaeus 1758)
Smallmouth Bass *Micropterus dolomieu* La Cepede 1802
Largemouth Bass *Micropterus salmoides* (La Cepede 1802)

Perch Family Percidae
Don Ruffe *Gymnocephalus acerina* (Gueldenstaedt 1775)
Balon's Ruffe *Gymnocephalus baloni* Holcik & Hensel 1974
Ruffe *Gymnocephalus cernuus* (Linnaeus 1758)
Striped Ruffe *Gymnocephalus schraetser* (Linnaeus 1758)
European Perch *Perca fluviatilis* Linnaeus 1758
Percarina *Percarina demidoffi* Nordmann 1840
Asprete *Romanichthys valsanicola* Dumitrescu, Banarescu & Stoica 1957
Pikeperch *Sander lucioperca* (Linnaeus 1758)
Sea Pikeperch *Sander marina* (Cuvier & Valenciennes 1828)
Volga Pikeperch *Sander volgensis* (Gmelin 1788)
Asper *Zingel asper* (Linnaeus 1758)
Streber *Zingel streber* (Siebold 1763)
Zingel *Zingel zingel* (Linnaeus 1766)

Cichlid Family Cichlidae
Chanchito *Cichlasoma facetum* (Jenyns 1842)

Blenny Family Blenniidae
Freshwater Blenny *Salaria fluviatilis* (Asso 1801)

Goby Family Gobiidae
Banded Tadpole Goby *Benthophiloides brauneri* Beling & Iljin 1927
Rough Tadpole Goby *Benthophilus granulosus* Kessler 1877
Caspian Tadpole Goby *Benthophilus macrocephalus* (Pallas 1787)
Stellate Tadpole Goby *Benthophilus stellatus* (Sauvage 1874)
Caspian Goby *Caspiosoma caspium* (Kessler 1877)
Economidis' Goby *Economidichthys pygmaeus* (Holly 1929)
Trichonis Goby *Economidichthys trichonis* Economidis & Miller 1990
Berg's Goby *Hyrcanogobius bergi* Iljin 1928
Caucasian Goby *Knipowitschia caucasica* (Kawrajsky 1899)
Korission Goby *Knipowitschia goerneri* Ahnelt 1991*
Longtail Goby *Knipowitschia longecaudata* (Kessler 1877)
Miller's Goby *Knipowitschia milleri* (Ahnelt & Bianco 1990)*
Mrakovcic's Goby *Knipowitschia mrakovcici* Miller 1991*
Panizza's Goby *Knipowitschia panizzae* (Verga 1841)
Orsini's Goby *Knipowitschia punctatissima* (Canestrini 1864)
Thessaly Goby *Knipowitschia thessala* (Vinciguerra 1921)*
Toad Goby *Mesogobius batrachocephalus* (Pallas 1814)
Ginger Goby *Neogobius cephalarges* Pallas 1811
Monkey Goby *Neogobius fluviatilis* (Pallas 1814)
Racer Goby *Neogobius gymnotrachelus* (Kessler 1857)
Bighead Goby *Neogobius kessleri* (Gunther 1861)
Round Goby *Neogobius melanostomus* (Pallas 1814)
Syrman Goby *Neogobius syrman* (Nordmann 1840)
Martens' Goby *Padogobius martensii* (Gunther 1861)
Italian Goby *Padogobius nigricans* (Canestrini 1867)
Canestrini's Goby *Pomatoschistus canestrinii* (Ninni 1883)
Common Goby *Pomatoschistus microps* (Kroyer 1838)
Tubenose Goby *Proterhorinus marmoratus* (Pallas 1814)
Relict Goby *Relictogobius kryzanovskii* Ptschelina 1939
Grass Goby *Zosterisessor ophiocephalus* (Pallas 1814)

Sleeper Family Eleotridae
Chinese Sleeper *Perccottus glehni* Dybowskii 1877*

Flatfish Family Pleuronectidae
Atlantic Flounder *Platichthys flesus* (Linnaeus 1758)
Arctic Flounder *Pleuronectes glacialis* Pallas 1776

Anatomy

Head

In fish, most of the obvious external sense organs are located on the head: a pair of eyes, the nostrils (normally paired) and often barbels which may vary in number, size and position according to species. That part of the head anterior to the mouth is normally termed the snout. The position of the mouth itself can vary: it may be terminal, superior or inferior. In a few species of fish the mouth is modified to form a sucker. Associated with the mouth are several bones of taxonomic importance (see fig. 2); several of these sometimes carry teeth which may be long or short, permanent or deciduous. In adult lampreys oral discs are present which have supra-oral and infra-oral areas bearing teeth (see page 62). The mouth opens into the pharynx and in some fish there are, at the back of this, bones specialised for chewing and crushing, known as pharyngeal bones; the structure of these is extremely important in the identification of Cyprinidae (fig. 2). A diagrammatic representation of the form and positioning of the bones in the mouth of a fish is shown in fig. 3.

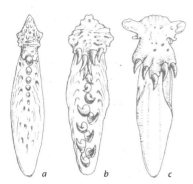

Fig. 2a Typical salmonid vomer bones:
a. Atlantic Salmon; b. Brown Trout;
c. Arctic Charr

Gills

Passing out from the sides of the pharynx are cavities leading past the main respiratory organs, the gills; together these form the branchial region (see fig. 4). Each gill consists of a strong supporting arch, on one side of which is a set of comb-like rakers, whose function is to prevent material passing from the mouth into the delicate blood-filled respiratory lamellae, which are aligned on the other side of the gill arch. There are normally four gills on either side, the passages between them leading to the outside of the body through the gill openings. In most species these are protected by a single bony gill cover on either side, known as the operculum.

Skin

The whole body is covered by skin; in most fish small bony plates known as scales lie within this, forming a protective but extremely flexible covering over most areas except the head, where protection is afforded by the head bones themselves. The number and structure of the scales varies greatly from species to species and they are often useful for identification purposes (see fig. 5). It should be noted, however, that in one species at least, the Common Carp *Cyprinus carpio*, there are cultivated varieties which have no scales at all (Leather Carp), or which have a few very large scales (Mirror Carp). Some other species of fish have no scales, and in others they are replaced by isolated bony scutes which project from the skin. Within the skin are pigment cells responsible for much of the coloration of the fish. Although colour patterning is a useful criterion for distinguishing various species of fish, it is one which should be regarded with caution, for even within a single species the colour can vary greatly with age, sex, season, time of day, emotional state, etc.

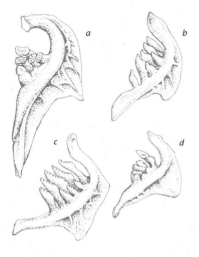

Fig. 2b Typical cyprinid pharyngeal bones:
a. Carp; b. Silver Bream; c. Rudd; d. Tench

To the fish biologist, scales present something more than just part of fish anatomy. Because of the structure of scales and the way they are laid down in the skin, each represents a permanent record of the growth of its bearer in much the same way as the growth rings in the trunks of woodland trees (see fig. 7). Thus, given a single good scale from a mature fish, a competent biologist can often identify the species, give its age in years, and say how often it has spawned. In migratory fish the biologist can often tell how long the specimen has spent in fresh water and how long in the sea, or how long in its nursery stream and how long in its present lake. In this way a detailed knowledge of scale anatomy is indispensable to any competent ichthyologist. Fig. 5 shows the general features of fish scales.

Only a few families of fish have no scales. In those which do there are normally differences between scales from different parts of the body – those from the head region, the lateral line and adjacent to the fins normally showing some modification of shape. Most keys refer to typical body scales (always the great majority) which are found above and below the lateral line between the head and the tail. The scales lie overlapping along the body of the fish like slates on a roof but, unlike slates, normally only a fraction of each scale is exposed to view. The width of the circuli (ring-like ridges on fish scales), and the number laid down, is related to the growth of the fish: wide circuli indicate good (usually summer) growth, narrow circuli indicate poor (usually winter) growth. Groups of narrow circuli are normally formed each winter and are called annuli.

Apart from scales, other bony parts of the body can also be used for determining age: among the most important of these are opercular bones, otoliths and fin rays.

Running along either side of the body in most fish is the lateral line; this is a long sensory canal just under the skin, but connected to the exterior by a series of pores. These often run through individual scales. The main function of the lateral line is sensory – the fine detection of various kinds of vibration through the water medium. Branches of the same system run on to the head region, where they terminate in sensory organs often found in grooves and cavities.

Fins

The typical arrangement of fins on a fish is shown in fig. 8. There are two sets of paired fins – the pectoral fins and the pelvic fins – both situated ventrally. These are equivalent to the fore and hind legs respectively of terrestrial vertebrates. On the back is a dorsal fin; this may occasionally consist of two distinct parts, or may be divided into two separate

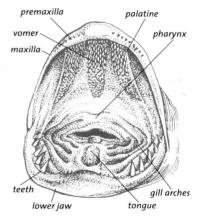

Fig. 3 Open mouth of Pike showing major bones and teeth

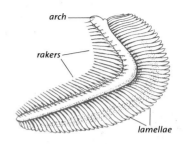

Fig. 4 Typical fish gill

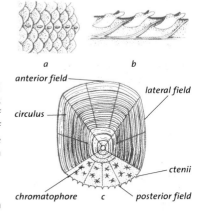

Fig. 5 Fish skin and scales:
a. side view of scales near lateral line:
b. section through lateral line showing its ducts passing through individual scales;
c. individual scale

fins, or may have the anterior of these represented by several isolated spines. Behind the dorsal fin, in Salmonidae and related families, is a small fleshy fin with no rays called the adipose fin. The portion of the body posterior to the anus is known as the caudal region. Ventrally, just behind the anus, this bears the anal fin, while posteriorly, behind where the body narrows to form the caudal peduncle, is the single caudal fin. The supporting structures of fins are known as rays; these may be branched or unbranched (when they are usually referred to as spiny or bony), and are often useful taxonomic characters.

Fig. 6
The Mirror Carp
is a variety of
Cyprinus carpio
that has been
cultivated to
have only a few
large scales

Internal organs

Internally there are a number of features which it is important to understand. As in other vertebrates the body of a fish is supported by a bony, but flexible, vertebral column. To this are linked the head and fins of the fish and the numerous blocks of muscle (often interspersed with fine bones) which run along either side of the body. This is that part of the body which gives propulsion, but it is also that most sought for food by man and other predators.

Below the vertebral column is the major body cavity of the fish, containing many of its vital organs (see fig. 9); these can be examined properly only by dissection and are normally removed entirely during gutting. The alimentary canal or gut starts at the pharynx, which leads into the oesophagus which opens, in turn, into the stomach. Food eaten by the fish is held for some time in the stomach before being passed into the intestine where it is digested; undigested materials continue into the rectum, from which they pass out through the anus as faeces. Most fish are able to control their buoyancy in the water by means of a swim-bladder which is situated above the gut, and which in some groups is connected to the oesophagus by a duct.

Associated with the gut is the liver, which is often used by man as food or as a source of rich oil. Above the gut on either side normally lie the sexual organs. These are relatively simple in

Fig. 7
Proportionate
growth:
the relationship
between annual
body length and
size of scale in a
cyprinid fish

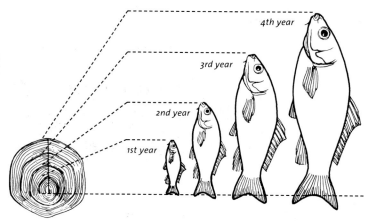

4th year

3rd year

2nd year

1st year

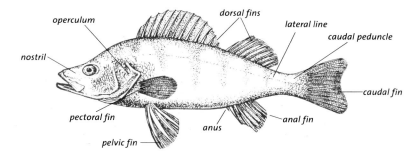

Fig. 8 Main external features of a typical fish (side view)

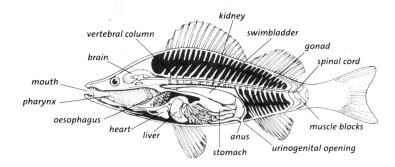

Fig. 9 Main internal organs of a typical fish (side view)

most fish and consist of two elongated bags opening to the exterior through a genital duct beside the anus. The sex of many fish can be established only by examination of these organs: the ovaries normally contain what are obviously globular eggs – often yellowish or orange, while the male testes are normally rather smooth and whitish. Immediately below the vertebral column in most fish lie the kidneys and the swim-bladder. The latter, by the adjustment of its volume through gases being released from, or absorbed into, the blood of the fish, allows the maintenance of neutral buoyancy. This means that the fish has to expend no effort in maintaining its vertical position in the water.

Below the head and just posterior to the gills lies the muscular heart, which is responsible for pumping blood through the gills to be oxygenated, and then through the body of the fish.

The fish's brain is well protected inside a bony capsule and, although much simpler than the equivalent structure in birds and mammals, is none the less a complex lobed organ (see fig. 10). At the back of the head, also encased in bone, are the inner ears or semicircular canals which are particularly important in maintaining balance. Within a chamber inside them is secreted a loose piece of calcium carbonate, known as an otolith. This, like the scales, grows in proportion to the size of the fish.

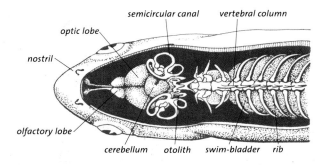

Fig. 10 Brain and balancing organs in a cyprinid fish (dorsal view)

Physiology

Digestion

Like all other animals, fish require food to live and grow. This food is always produced in the first instance by plants, but may come to the fish concerned indirectly via an invertebrate or a more complex food chain. By observation in the field and experiments in the laboratory, fish biologists have discovered much concerning the feeding behaviour and types of food eaten, as well as the nutritional value of different foods. Much of this work is very recent and related to developments in fish farming discussed later.

A brief description has already been given of the mouth and gut of a fish. Not only is the mouth adapted to the type of food eaten but the gut is also often modified in this respect. Thus the oesophagus and stomach are very distensible in carnivorous fish, allowing them to swallow whole fish which are very large relative to their own size. The main purpose of the digestive system is to break down foods into soluble materials which can be absorbed through the gut wall and used by the fish for growth and metabolism. After food has been swallowed it is acted upon by various enzymes secreted by the gut and organs associated with it. Important among these is the liver which secretes bile, an important aid to digestion, and which also acts as a storage and processing organ for food after it has been absorbed.

Food material is moved down the gut by peristaltic waves of contraction of muscles in the gut wall. Although in some fish a little absorption takes place in the stomach, it is in the intestine that most food material passes into the bloodstream as soluble fats, proteins and carbohydrates. Once food has been digested it can be used to provide energy for movement, materials for replacing or regenerating old cells, or for growth. Since the body temperature of a fish is controlled by that of its environment, all these processes of metabolism are fast at high temperatures and slow at low temperatures. The efficiency of fish in converting food is variable but it is known in hatchery conditions, for instance, that Rainbow Trout have a conversion efficiency of about 3.5; i.e. for every 3.5 kg of food eaten they increase in weight by 1 kg.

Growth

The growth of fish is closely concerned with the quality and quantity of food taken in, although many other factors (space, temperature, etc.) are involved too. One of the outstanding features of fish is their phenomenal plasticity as far as growth is concerned. When food and other conditions are suitable many fish are able to grow at very fast rates, but in adverse conditions, often with no food for long periods, they are able to maintain themselves with no growth at all. This is in considerable contrast to most birds and mammals which are much less flexible in this respect, and usually die after a relatively short period without food. Unlike birds and mammals, fish continue to grow throughout their lives and do not stop on becoming sexually mature. Many small species of fish – especially in tropical zones, live for only 1 or 2 years, but some large species may live for 30 years or more.

Fig. 11 Main blood vessels and direction of flow in a typical fish (side view)

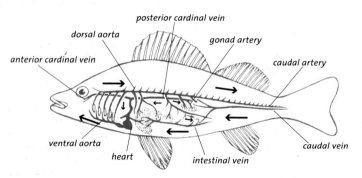

Circulation

The circulatory system of a fish is one of the major aspects of its physiology, linking digestion, respiration and excretion. Its main function is to carry oxygen and carbon dioxide, cell wastes and products of excretion, and minerals and dissolved foods through the body. The blood of fish, as in other vertebrates, consists of a fluid plasma in which are dissolved various materials, and in which the solid blood cells are carried. These blood cells are of two types: white lymphocytes and leucocytes, and red erythrocytes. The red haemoglobin in the latter greatly aids oxygen transport in the blood. The actual amount of blood present in the body of a fish is quite low, and is usually only about 2–3 per cent of the body weight (it is more than 6 per cent in mammals).

The circulatory system of fish is relatively simple, consisting basically of a continuous tubular system of heart, arteries, capillaries and veins (see fig. 11). The heart (which contains one-way valves) pumps the blood forward into the gills where it passes through fine capillaries in close contact with the water outside. The blood then collects in vessels which transport it to the various tissues of the body where it again passes through capillary systems. The blood then passes into veins and back directly to the heart, although some flows through the liver and the kidneys where it passes through yet another capillary system before moving to the heart. Blood pressure is at its highest on leaving the heart, but drops considerably after passing through each capillary system and is quite low by the time it passes through the final major veins into the heart.

Respiration

Respiration in fish, as in other animals, is concerned with the intake of oxygen and the elimination of waste carbon dioxide – one of the main products of cellular activity. The gills of fish are equivalent to the lungs of many terrestrial animals, and represent the site where oxygen from the water in which the fish is living enters the bloodstream, and carbon dioxide leaves it. In a number of adult fish and in many fish fry, some respiration takes place through the skin. It will be obvious that respiration in fish is closely linked with the circulatory system – particularly in the area of the gills.

There are two main types of gills found among fish. In the pouch-like gills of the lampreys (and other Agnatha) each pouch has an internal opening to the pharynx and an external opening to the water. The branchial gills of higher fish, however, are carried on arches on either side of the pharynx and connect individually to the outside through a series of gill slits (as in sharks and rays) or through a single opening protected and controlled by the gill cover or operculum. In a few unusual types of fish, particularly those which may have to withstand drought from time to time, there are specialised organs of respiration involving the gut or swim-bladder. In all gill systems the main objective is the same: to allow maximum contact of blood in the gill capillaries (enclosed in thin epithelial tissue) with water moving past the gills.

The water moving over the gills is kept in constant unidirectional motion by the fish. When the fish's mouth opens, water is sucked inside and fills the whole of the mouth and buccal cavity, on either side of which are the gills. The mouth then closes and water is forced between the gills and out past the opercula which are opened at this point. Thus the mouth and opercula are in constant alternate motion at a rate dependent on the oxygen requirements of the fish. The flow of water is always one way, except in the case of lampreys where, when use of the sucker prevents entry of water, this passes both in and out of the gills by the pumping action of the buccal cavity. The respiratory system of fish is made even more efficient by the fact that the flow of blood within the gills, and of water surrounding the gills, is in opposite directions, blood passing from the heart forwards, water passing from the mouth backwards. Counterflows of this nature are the most efficient methods of transferring materials from one solution to another.

Most oxygen taken up in the blood of fish is carried in red blood cells. This allows far larger amounts to be carried than are contained in the same volume of water. As oxygen is taken up, so carbon dioxide is released, and when the blood finally passes from the gills to the tissues it is rich in the former but deficient in the latter.

Excretion

A further function of the gills of fish which should not be forgotten is that concerned with the uptake and excretion of various salts. Fish excrete their body wastes in various ways, partly through the gills, partly into the gut to pass out with the faeces, but mainly, as in other vertebrates, through the kidneys, from which the waste products pass to the exterior through special ducts. In fish the kidneys have the dual function of eliminating body wastes and helping to control the water-salt balance. Fish blood is in a constant and dynamic equilibrium with its surroundings, in other words it can make changes in response to changing external salt concentration. This is a result of the process known as osmosis, in which water in solutions separated by a permeable membrane will pass from the more dilute solution into the more concentrated until they are at the same concentration. Thus in sea water, which is more concentrated than blood, water tends to pass out from the fish mainly through the gills, and fish have to drink water regularly to compensate, while in fresh water, water tends to pass in through the gills and must constantly be discharged via the kidneys to compensate. The ability to control the quality of the blood in relation to outside salt concentration is the main factor preventing freshwater fish from living in the sea and vice versa. Fish which are able to pass from sea water to fresh and back have special excretory abilities.

The kidneys of fish are made up of numerous small tubules. These act, under the pressure of the blood, as minute filters which take out various, mainly nitrogenous, salts from the blood and pass these, together with excess water, to the outside via the renal ducts. These open very close to the anus of the fish.

Behaviour

The behaviour of fish, like that of other vertebrates, is made up of two components – instinct and learning. Much of the life of the fish is dominated by the former, but the learning process should not be underrated, as a visit to a modern fish farm will show. Here, fish such as Rainbow Trout learn quickly to assemble at particular places to be fed, or even to feed themselves from automatic dispensers which release food when the fish presses on a lever.

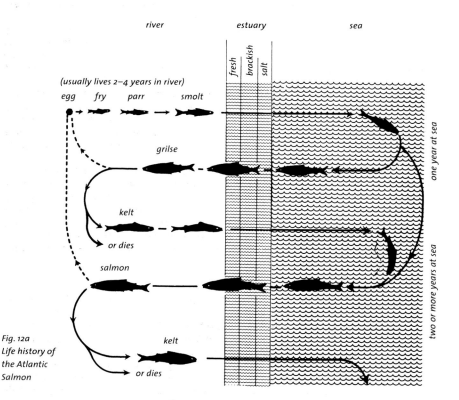

Fig. 12a
Life history of
the Atlantic
Salmon

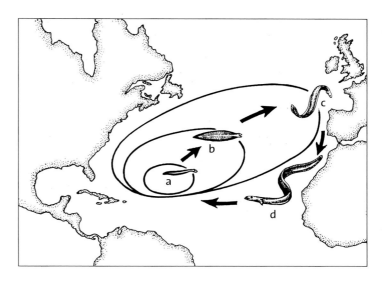

Fig. 12b Life history of the Eel: a. young larva leaving the Sargasso area where it was spawned; b. maturing larva (leptocephalus) drifting across the Atlantic; c. elver about to enter fresh water on European coast; d. adult Eel returning to the Sargasso to spawn

Shoaling

One of the simplest kinds of behaviour is that shown by shoaling species. Many species of fish are solitary virtually all their lives, except to meet up with members of the opposite sex for spawning purposes, while others spend most of their lives in company with members of their own species, forming shoals which may often number many thousands. Most of the purely solitary species are predatory and often large: the Pike is a good example. Other than on their spawning grounds, members of this species are rarely found together in any numbers. One of the reasons for this is that large Pike regularly eat smaller Pike, and in certain situations where suitable food species are rare or absent they eat little else.

Shoaling species on the other hand tend to be smaller, herbivorous, or more commonly omnivorous, fish which keep together in packs, the density of which usually depend on the activity in which they are engaged. Shoaling fish often, but by no means always, tend to be silvery in colour, living in open areas of water. The Roach is a good example. Almost immediately after hatching the young start to congregate and move about as one unit. This unit is at its most dispersed at night or sometimes when feeding, but when moving about – especially rapidly if danger threatens – a very tight pack is formed. These shoals may break up to form smaller units or join together to form larger ones, but many of the fish may remain together most of their lives. Spawning is likewise a shoaling activity, sometimes preceded by a very obvious migration in which enormous shoals move to one part of a lake or migrate upstream into rivers to lay their eggs.

The function of shoaling behaviour is usually thought to be one of mutual protection and advantage. It is much more difficult for predators to approach without being seen, while a new source of food discovered by one member of the shoal is very soon likely to be engaging the attention of most of its members. Sometimes fish are found together in considerable numbers and are called shoals, but the congregation is one of chance and there is no interplay or reaction between fish as in a true shoal. Thus during winter certain fish, for instance Sturgeon, may congregate in deep holes in lakes or rivers and remain together in a rather torpid state for many weeks.

Migrating

This aspect of fish behaviour has received considerable attention from fish biologists and others. Many more fish undertake migrations than is commonly supposed, but these are often of a local nature and do not involve the spectacular distances undertaken, or location problems solved, by the better-known species. A good example of small-scale migration is found in Brown Trout, populations of which are found in many thousands of lakes all over northern and highland Europe. The adult Brown Trout in these lakes are territorial and often occupy one small area of the lake for most of the year. In the autumn, however, they move to the mouths of streams entering the lake and migrate upstream to the spawning grounds.

Normally Brown Trout only migrate into the stream in which they themselves were hatched.

In both Atlantic and Pacific Salmon a similar pattern of migratory behaviour is exhibited, but the distances travelled and obstacles surmounted are spectacular. It is now known that many Atlantic Salmon reach maturity in the seas off Greenland. In order to reach the European (or North American) coast these fish have to migrate for many hundreds of kilometres and then locate the mouth of the river from which they originated (see fig. 12a). They then have to migrate through the estuary with its violent changes of salinity (and in modern times not insignificant pollution) and upstream to the headwaters, sometimes leaping waterfalls over 2 m (6.5 ft) in height. In contrast, the migration of the few adult fish which live to return to the sea, or of the descending smolts, seems a much simpler and more passive affair, but in fact many of the same problems are involved. Although much studied, some of the patterns of behaviour responsible for the migratory abilities of these fish are still not clearly understood. In general it is felt that orientation over long distances, say in the sea, is related to physical features such as water currents, but that in location and selection of parent rivers, chemistry is more important.

Migratory fish species such as Atlantic Salmon which move into fresh water to spawn but then pass down to the sea to grow to maturity are called *anadromous*. Sturgeon, Sea Lamprey, Allis Shad and several other species come into this category. *Catadromous* species, on the other hand, show the opposite type of behaviour, growing up in fresh water but migrating downstream to the sea to spawn. Eels and some populations of Flounders are good examples of this category.

Breeding

The behavioural aspects of fish biology reach their most complicated and bizarre at breeding time. In many species only then is it possible to distinguish externally between the sexes, sometimes simply because the female is very plump and swollen with eggs, but more often because the male has become brightly coloured or has developed tubercles. These are small white lumps which appear on the head and sometimes the fins and bodies of sexually mature fish during the spawning season. They are most common among Cyprinidae, but occur in some other families too (e.g. Coregonidae). Many types of behaviour are shown by a single species of fish during the breeding season; apart from the spawning act itself, these range from aggression towards males of the same species or predators of the eggs and young, through complicated nest-building activities, to fanning the eggs in the nest and actively 'herding' the young fish until they can fend for themselves.

Fig. 13 Spawning behaviour in the Bitterling

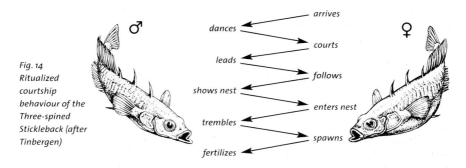

Fig. 14
Ritualized
courtship
behaviour of the
Three-spined
Stickleback (after
Tinbergen)

The Cyprinidae as a whole show relatively little in the way of sophisticated spawning behaviour, although a distinct exception to this is the Bitterling. This is a small, unobtrusive fish which lives in ponds, canals and slow-flowing rivers; for much of the year the sexes are alike. At the beginning of the breeding season in late spring, however, the male becomes much more colourful and exhibits beautiful blues and reds on the body and fins. The female at this time grows a long, thin, flexible tube from the genital opening, which is an ovipositor. A pair of fish establish a territory in an area within which is a freshwater mussel. After preliminary courtship behaviour, the pair approach the mussel and the female, using her ovipositor, lays one or more eggs actually inside the mussel via its inhalant siphon; these are fertilized instantly by the male (see fig. 13). The process is repeated until all the eggs are laid. Inside the mussel the eggs are oxygenated by the water currents created by the mussel until they hatch and the fry swim outside. After the spawning season the adults lose both their external sexual characteristics and their interest in mussels.

One of the best studied examples of sexual behaviour in the animal kingdom is that of the common Three-spined Stickleback. Outside the breeding season the sexes are very similar – usually a dull mottled grey-green – but in spring the male develops a bright red throat and belly, and iridescent blue-green eyes and sides. The female becomes silvery and very plump. Always a rather aggressive species, the male becomes particularly so at this time, establishing a territory and driving away intruders of all kinds. Within this area he starts nest building, clearing a shallow depression in which is built a tubular nest made up of fine pieces of plant material. Any ripe females in the area are actively courted and led to the nest where they lay their eggs, which are then fertilized by the male. Usually the procedure is repeated with two or three females. These are driven away following spawning and the male defends the nest, keeping it clean and periodically fanning the eggs to keep them oxygenated. Even after the eggs have hatched the father protects the young for some days until they can swim well and fend for themselves. After the breeding season the fish lose their colours and interest in territories.

Thorough and painstaking analyses of this apparently complex behaviour have shown that it is completely stereotyped and each part of it will only take place when the fish has been 'triggered' by some release mechanism. The spawning pattern is shown diagrammatically in fig. 14. The male will only perform the characteristic zig-zag dance in the presence of a ripe female. The two significant features of the female to be her silver colour and her plumpness, for the male can be persuaded to perform before a crude model as long as it is silver and has a large belly. Each display or movement by one sex releases the next movement in the other until the whole sequence is carried through or fails to develop through lack of appropriate response.

Feeding

The behaviour of fish during feeding is also of extreme interest to biologists, anglers and others. It is at this time that fish exhibit their greatest potential for learning, and the opportunist nature of many species has led to their considerable success in some waters. The feeding patterns of some species, notably predators such as pikeperch on the one hand, and filter feeders such as lamprey larvae on the other, are very instinctive and stereotyped. Other fish are more adaptable. Thus Brown Trout in a river may be feeding actively on benthic invertebrates on the bottom one day, but ignoring these completely the next day to feed on mayflies emerging near the edge. On the third day both these sources may not appear in the diet, and instead it will consist of terrestrial insects blown on to the water surface from nearby trees. In changing circumstances like these, opportunist fish species tend to choose the food source which is most easily available for the least expenditure of energy.

Development

Sperm and eggs

Almost all European fish are oviparous, i.e. the sperm and eggs are ejected close together in the water, and after fertilisation the egg undergoes development quite independent of its parents, although they may protect it and keep it clean, etc. An exception to this system is the introduced Mosquito Fish, where the anal fin of the male is modified to form an elongate penis. This is used to fertilise the eggs inside the female, where they remain protected until they hatch and the young are born alive. They do not, however, receive any food materials from the female subsequent to their fertilisation.

In all the other species in Europe the reproductive systems are essentially of one type, consisting of paired gonads (ovaries in the female, testes in the male) and their ducts leading to the exterior. In the testes a process known as spermatogenesis gives rise to the development of specialised sex cells known as sperm. These carry the hereditary characters of the father, and in order to ensure successful fertilisation they are produced in huge numbers. Each sperm has a long whip-like tail which enables it to swim about in the seminal fluid secreted by the sperm ducts and subsequently in the water.

Oogenesis is the process in the female equivalent to spermatogenesis, and leads to the development of varying numbers of eggs within the ovary of the female. Like sperm, each egg carries hereditary characteristics of its parent

Fig. 15
Ventral view of dissections showing the typical cycle of ovary development in a fish

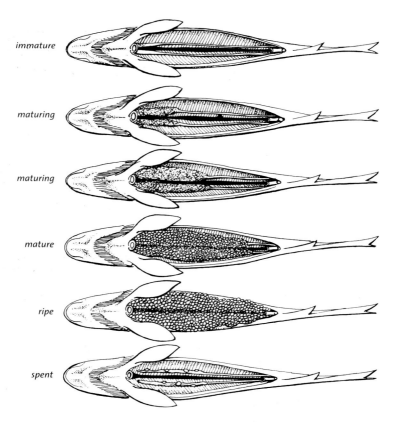

immature

maturing

maturing

mature

ripe

spent

Fig. 16 Early development in the Common Whitefish (after Slack et al.): a–c = growth of larva and absorption of yolk; d = hatching; e = early larva with yolk sac; f = late larva; g = fry

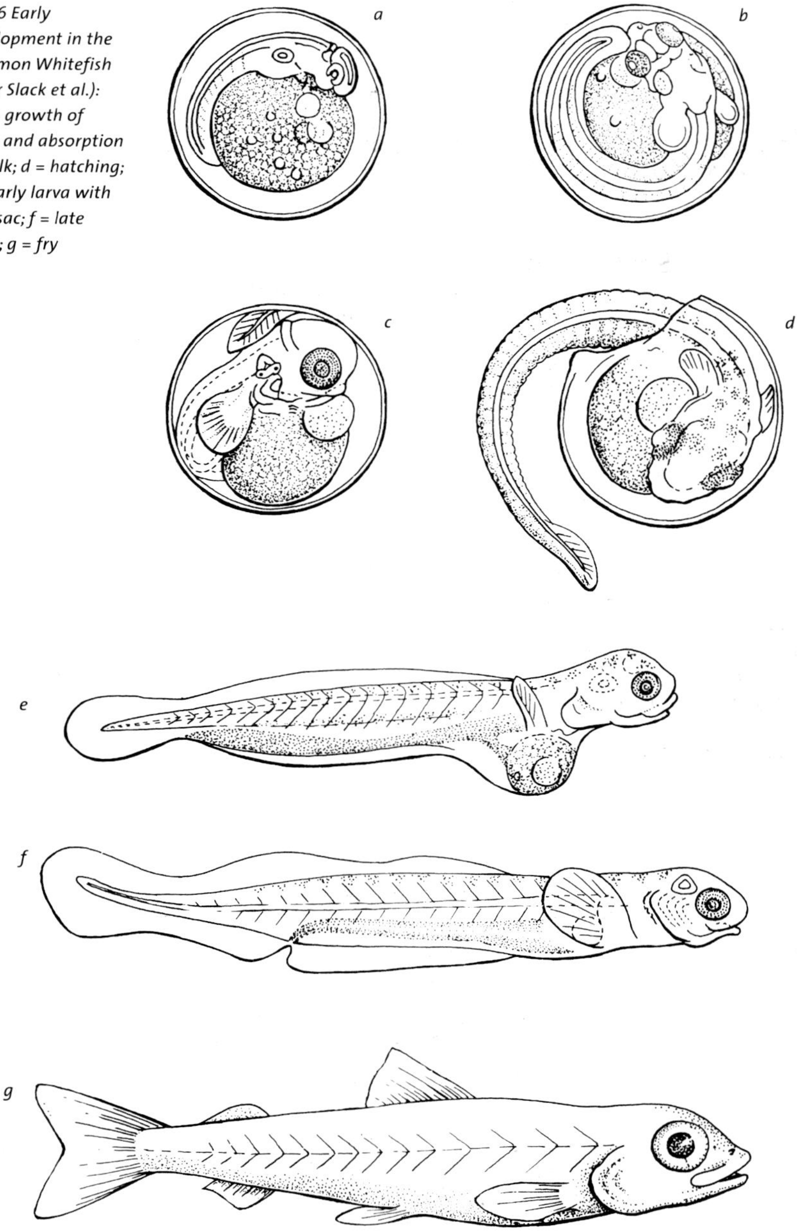

but the cell itself is very much larger, having been provided during development with large quantities of yolk and fat. The number of eggs produced by a female fish generally increases with her size and varies tremendously among species. Thus small fish like Three-spined Sticklebacks and Bitterling, whose eggs and early fry are afforded considerable protection after laying, lay relatively few (50–100) eggs at each spawning. Larger species, however, whose eggs are shed into the water and given no further protection lay very large numbers each year (up to 1 million in the case of Flounder). The number of eggs laid by a female is referred to as her fecundity.

Reproduction in almost all European fish is a cyclic process, usually related to the seasons of the year. It is controlled by reproductive hormones whose secretions are in turn dependent on outside factors such as temperature and day length. The gonads, particularly the ovaries, may undergo tremendous changes in size and appearance during the year, starting at their smallest just after the spawning season and increasing to a maximum just before the next season, when the sexual products are shed (see fig. 15). Although the actual spawning period is normally relatively short for any population of a species (usually a few weeks) the time of year at which different species spawn can vary tremendously. Thus in Europe some fish are spawning in each month of the year.

After spawning, the eggs of different fish species may find themselves in very varying situations. Many species construct a nest to give the eggs some protection during development. This 'nest' may be simply an open depression in the substrate, as in the introduced Pumpkinseed; a similar hole in which the eggs are laid but then covered by additional substrate, as in the Atlantic Salmon and Brown Trout; a space cleared out underneath a rock (Bullhead); or a much more complicated structure created from pieces of weed, as in the Three-spined Stickleback. Some species protect the eggs and often the young fish in these nests (Pumpkinseed), while others leave them immediately after spawning (Atlantic Salmon).

Eggs which are spawned without the protection of a nest of some kind can be laid in various ways. A few species, Perch for example, lay long strings of eggs which tangle up among vegetation. Many others lay adhesive eggs which may stick to stones (Atlantic Sturgeon), to plants (Common Carp and Northern Pike), or to sand and plant roots (Gudgeon and Spined Loach). The spawning of Bitterling inside mussels has already been discussed in detail. Finally, other species spawn in the water, and the eggs may float (Golden Grey Mullet), or sink to the bottom and lie loose there (Vendace).

The time taken for egg development varies tremendously among species, and within species is very dependent on temperature. Usually the eggs of fish which spawn in spring – and especially in summer – hatch quickly. The incubation period in Carp lasts only 3–5 days. In fish like the Atlantic Salmon which spawn in autumn and winter on the other hand, the eggs may take more than 150 days to hatch.

During incubation the egg itself naturally undergoes profound changes. Most of its volume is occupied by the yolk; the cell resulting from the union of the sperm and the ovum being very small. This cell undergoes a process known as cleavage to give 2 cells, then 4, then 8, 16, 32 and so on, and each group of cells starts to grow and differentiate into a part of the embryo. The yolk is gradually used up during this process and eventually the egg consists of a spherical membrane, inside which is curled up a small embryo fish. The eyes and the constantly beating heart can be seen quite clearly at this stage.

After hatching, the young fish may start swimming immediately and continue to do so for more or less the rest of its life (for instance Common Whitefish) or, more commonly, it will rest in a protected place for some time until the remains of the yolk sac are fully absorbed. Atlantic Salmon remain among gravel in their nest during this period, while Pike have a temporary adhesive organ by means of which they hang, resting, attached to vegetation. At some time during this period most species make a swift visit to the water surface to fill the swim-bladder initially.

Larvae and juveniles

The very young stages of fish are given a variety of names, the meaning of which is often rather imprecise. In general, the term larva is used to describe the stage from hatching until the fish is a miniature adult (see fig. 16). This may take only a few days in carp, but up to several years in European Eels and Sea Lampreys. In some fish, including the two just mentioned, the two life stages are so different that they were originally described by scientists as different species. The larval stage itself is sometimes divided into prolarval, where the yolk sac is present, and postlarval, when it has disappeared. In general, larvae are characterised by transparency, absence of scales, presence of large pigment cells and embryonic, undifferentiated fins. Beyond the larval stage, young fish tend to look very much more like their adults although many features, particularly coloration and sexual differences, are not evident until full maturity is reached. As indicated elsewhere, most identification keys, including the one in this book, refer to mature specimens of the species concerned and some difficulty may be met if only young fish are available. If it is necessary to identify larval specimens, specialist keys must be used.

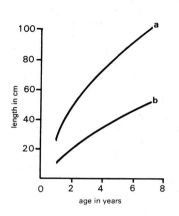

Fig. 17 Fast and slow growth in the Pike: a. Lough Rea, Ireland; b. Dubh Lochan, Scotland

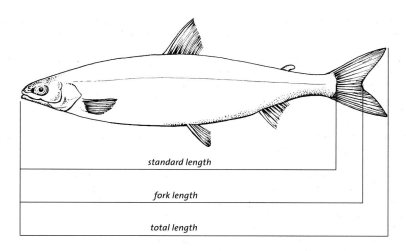

Fig. 18 Methods used for measuring the length of a fish

Age and growth

The age at which fish become mature (i.e. are able to reproduce) varies very much among different species and even among different populations of the same species. In general, sexual maturity is related to size, which in turn is dependent on growth. Thus fish in fast-growing populations tend to become mature earlier than those in slow-growing populations. Both food and temperature affect growth; fish at higher latitudes mature later than those at lower ones. In Finland, Roach do not mature until they are 5–6 years old, whereas in southern Europe the same species may be mature at 2–3 years.

Small species tend to mature and die early. The Stellate Tadpole Goby, for instance, matures and dies within 1 year, whereas the Beluga does not mature until it is 15–20 years old and may live for 10 or more years beyond that. Some fish spawn only once; the Atlantic Salmon spends 2–6 years in fresh water and a further 1–2 years in the sea before it is mature, but very few fish live beyond this to spawn a second time. In the case of the Pink Salmon and the European Eel there appears to be complete mortality after the first spawning. Other species such as Brown Trout, may spawn several times during their lives.

As previously mentioned, the growth of most European fish species takes place in annual spurts – usually during the warmer months of the year. These seasonal variations lead to physical differences in scales and various bones, which are of great value to fishery biologists in determining age and growth. Growth itself is controlled by many factors, among which the most important are food, temperature and genetic constitution of the individual concerned. An understanding of these and other factors is essential to the successful management of fish stocks, particularly in fish farms and small closed sport fisheries where success may be determined by the ability of the fish stock to achieve optimum growth (see fig. 17).

It has been shown that, in both benthic and pelagic feeding, fish growth is related to the amount of food available. Thus populations of Common Carp in ponds with high densities of invertebrates grow faster than those in ponds with low densities. The growth of the latter fish may be improved either by decreasing the number of fish, or increasing the density of the benthos (the animals and plants living in, or on, the bottom of the habitat). Changes in the quality and quantity of food can very quickly affect growth rates. In Atlantic Salmon, for instance, a dramatic increase in growth takes place immediately after migration from fresh water to the sea. For their first few years Sea Trout and Brown Trout may live together in streams and grow at identical rates. As soon as Sea Trout move into salt water, however, they start to grow at a very much faster rate than Brown Trout which have remained in fresh water. Growth is usually measured by increase in weight or in length. There are three different ways of measuring length in fish (see fig. 18), fork length being the most reliable and the measure in commonest use.

Ecology

Ecology is the study of animals and plants in relation to their environment. In the case of any particular species of fish this means understanding not only its physical and chemical surroundings, but also its biological surroundings – vegetation, predators and prey. The whole basis of speciation in the biological world is related to the fact, that over generations, species of animals and plants develop differently from one another as a result of one type being particularly successful in at least one environment. This implies that, although living together in the same body of water, different species tend to exploit different parts of it.

Habitat

Physically, fresh waters may be divided into two major categories – running and standing waters. The obvious primary distinction between these is the constant unidirectional flow of water in the former, and its absence in the latter. Each has different size categories. Running waters, for instance, range from small trickles and burns through streams to large rivers. Canals are an important type of man-made running water. Standing waters range from small (often temporary) pools, through ponds to large lakes (see fig. 19). The most important type of man-made systems in this category are reservoirs, which in some areas are the dominant type of water body.

Superimposed on the size and other physical attributes of a water (its altitude, geology, etc.) is its chemistry. Apart from the influence of salinity in waters in coastal areas, the chemical nature of a water depends on the geology of its catchment. Broadly speaking, base-rich rocks and soils (most

limestones, some sandstones, etc.) give rise to chemically rich water while base-poor rocks (granites and other igneous rocks) are associated with poor waters. Often the former are found in lowland areas and the latter in highland areas. The influence of man is important here, and the forms of agriculture carried out in a catchment can substantially affect the chemical quality of the water there. Lowland waters tend to be affected more than highland waters in this way.

Biologically speaking, two major types of water are recognised by freshwater biologists. Eutrophic waters are usually rather turbid, with chemically rich water

Fig. 19
Top: Lowland lake – favoured by cyprinids
Left: Upland lake – favoured by salmonids

Fig. 20 A stony upland stream – the favoured nursery habitat of many salmonid species

characterised by high pH and alkalinity. The invertebrate fauna is made up of large numbers of worms, leeches, snails, mussels and shrimps, while coarse fish, especially Cyprinidae, are the dominant vertebrates. In *oligotrophic* waters on the other hand, the clear water is low in chemicals and is typically slightly acid with a low to neutral pH. The invertebrate fauna is dominated by insects, especially stoneflies, mayflies and midge larvae, while Salmonidae are the major fish found. A third type of water, known as *dystrophic*, is found in some areas. This is characterised by high quantities of humic acids derived from peat which stain the water brownish. Few fish are found here.

Although waters can be classified into these various types there are often intermediate categories, and indeed a single system may be divided in several different ways according to which part is examined. This is particularly true in the case of large rivers. In the upper reaches of a typical river the main substrate is often peat near the source, followed by bare rock and boulders further downstream. The flow is very small (often not more than trickle in dry weather) and the gradient is steep. The dissolved salt content of the water and the average temperature are low, and higher plants are represented by mosses. The dominant invertebrates are stoneflies and very few fish are found. Those that do occur are usually only Brown Trout, but sometimes Bullheads and a few other highland species are also present.

In the middle reaches of the river the volume increases and the main substrate gradually changes from boulders to stones, and thereafter from stones to gravel. The gradient is now much less steep, and the dissolved salt content of the water is higher than upstream. A number of higher plants occur here, and a variety of invertebrates, particularly mayflies and caddisflies. A wider variety of fish occurs, often including Brown Trout, Bullhead, Grayling, Minnows and Stone Loach. In the lower reaches of the river the substrate of coarse gravel gives way to sand and eventually to fine silt near the mouth. Here the river is large, and the gradient slight, while the dissolved salt content is relatively high. The river is flowing at low altitudes and the average water temperature is much higher than near the source. A wide variety of invertebrates and plants occurs here, some typical of rivers, others commonly also found in ponds and lakes. The fish fauna is varied and includes many cyprinid fish such as Bream, Tench, Roach and Carp, as well as Perch, Pike, Eels and many others.

Near the sea the river opens out into a broad estuary and the ecological situation is dominated by the regular presence of salt, or at least brackish, water. Many of the fish found upstream are immediately restricted here, but a few are tolerant (e.g. Three-spined Sticklebacks and Dace), and many others pass through as they migrate to or from fresh water. Such fish include Sea Lamprey, European Eel, Atlantic Salmon and European Flounders. Typical estuarine species are Twaite Shad, Smelt and Thick-lipped Grey Mullet.

Zones characterised by different fish species must not be treated too seriously for they are obviously dependent on the species concerned being present in the river system. Thus most rivers in the north of Scotland have only two non-migratory freshwater species (Brown Trout and Three-spined Sticklebacks), whereas rivers in the south of Europe may have twenty to thirty species. Schemes of classification developed by ecologists for one area are often transplanted to others with most curious effects.

Food

The food available to fish is obviously a very important part of their environment, and indeed one on which the survival of each population depends (see fig. 21). Many fish are much more opportunist in their feeding behaviour than fishermen or scientists give them credit for, and although their diet may be virtually 100 per cent one type of food on one day, it may be quite different the next. In broad terms, however, fish may be divided into categories according to the type of food they eat, although it must always be remembered that the diet of young fish usually differs from that of adults.

A few fish feed on fine organic detritus and microscopic organisms on the beds of rivers. Lamprey larvae feed wholly on such material which is carried into the mouth by currents and filtered off in the pharynx. Zooplankton, those small transparent invertebrates which spend all their lives floating free in the water, are an important source of food to many young fish and some adults, notably Arctic Charr and Whitefish. Benthic invertebrates are a major source of food for fish and many species are highly modified (in the possession of barbels, etc.) for locating these small animals and grubbing them out of the bottom. Good examples of these are Atlantic Sturgeon, Common Carp, and Stone Loach. Plants are eaten in large quantities by very few fish in Europe, although Rudd and a few others eat them as part of an omnivorous diet. A number of fish are piscivorous and feed on other fish. Good examples of these are Northern Pike and Pikeperch. Such fish are often cannibalistic and can survive perfectly well in the absence of any other fish species. Thus several cases of pure Pike populations are known, where the young fish feed on invertebrates, while they themselves form the food of the adult fish.

The transfer of material in this way from one type of organism to another, which is then eaten by a third, is called a food chain or food web. The amount of organic material accumulated at each stage is referred to as production, and is most often measured as weight, or less commonly, but more precisely, as units of energy (such as calories or joules). At each stage in the chain there is considerable loss of energy due to metabolism and other processes, and usually only between 10 and 30 per cent is transferred to the next level.

The basis of most food chains in nature is the energy supplied by the sun. This is utilised by green plants (during the process known as photosynthesis) to elaborate more plant tissue from carbon dioxide, nutrient salts and water. These plants (called the primary producers) are then eaten by herbivorous animals (called the secondary producers) such as insect larvae, which in turn become the food of carnivorous fish (the tertiary producers) such as trout. Food chains may be quite simple, for example: diatom – Rudd – Northern Pike, or more complex, for example: diatom – water flea – water beetle – Brown Trout – Northern Pike – Osprey.

Parasites and predators

Parasites and predators are both major influences on fish populations. Fish are hosts to a wide range of parasites, including Viruses, Bacteria, Fungi, Protozoa, Nematoda (roundworms), Cestoda (tapeworms), Trematoda (flukes), Hirudinea (leeches) and Crustacea (lice). Various members of these groups may be ectoparasites on the skin, fins or gills of fish, or endoparasites in the gut, muscle, liver or other internal organs. In the wild, most fish harbour several species of parasite – often tens to hundreds of each species. It is surprising in some ways how well they are able to cope with such loads, but it must be remembered that it is not in their interests for most parasites to kill their hosts, unless they require to disable or kill the host so that it is eaten by the next host required to complete its life cycle – usually a bird (e.g. Heron) or mammal (e.g. Otter). Many parasites require three of four hosts for their full life cycle and a common sequence might be: water flea – fish – bird.

However, in the unnatural surroundings of fish farms, where fish are crowded and stressed, some parasites can become a major problem and major outbreaks of viral and bacterial diseases and epidemics of parasites are common. The ways of coping with this problem vary from destroying all the fish in the farm, sterilising the cages and starting again, to treating the fish with antibiotics or other medication. In the hundreds of sea cage farms off the north west coast of Europe, the Sea Louse *Lepeophtheirus salmonis* is a major parasite of Atlantic Salmon and millions of pounds have been spent controlling this parasite by giving the fish baths of dangerous chemicals and in researching other methods of control. Though the parasite originally infects the fish in cages from the wild, the build up of sea lice in the cages and in the sea around the cages is substantial, so much so that the recent decline in Sea Trout stocks in these areas has been blamed on the heavy parasite load picked up when the Sea Trout swim in the vicinity of salmon farms.

Just as fish have been transferred from other continents and established in Europe, so have they brought their own parasites with them and several of the parasites now infecting fish populations in Europe are aliens from abroad. A good example is the gill fluke, *Urocleidus principalis*, which was found to have been introduced with its host, the Largemouth Bass, and is now established in lakes in southern England. Previously, this parasite was unknown outside its native North America. Even within Europe there have been notable transfers of parasites along with their fish hosts, one of the most important in recent years being the transfer of the skin fluke *Gyrodactylus salaris* from resistant stocks of Atlantic Salmon in Sweden to wild stocks in Norway which have subsequently been devastated by this small parasite. So much so, that the only way in which the Norwegian authorities have been able to eradicate it is to poison out all the fish in whole river systems.

Like most parasites, fish predators are normally in balance with their prey and are rarely a threat to fish populations in the wild. Thus birds such as Heron *Ardea cinerea*, Goosander

Mergus merganser and Cormorant *Phalacrocorax carbo*, and mammals like Otter *Lutra lutra* and Mink *Mustela lutreola* are common inhabitants of wetlands, where they feed largely on fish and are rarely a problem. Difficulties usually arise only where humans are involved, where anglers erroneously perceive such predators as threatening their sport, or fish farmers are surprised to have to take action against predators coming in for an easy meal. Many of these piscivorous birds and mammals have suffered tremendous persecution in the past and some species have become endangered as a result. Fortunately, Herons, Otters and many other fish-eating animals are now protected by European and local national laws and populations are recovering. Many people forget that small fish (larvae and fry) have a wide range of insect predators in the form of water boatmen (e.g. *Notonecta*), water beetles (e.g. *Dytiscus*), dragonfly larvae (e.g. *Aeshna*) and others – but fortunately anglers have never taken action against these creatures!

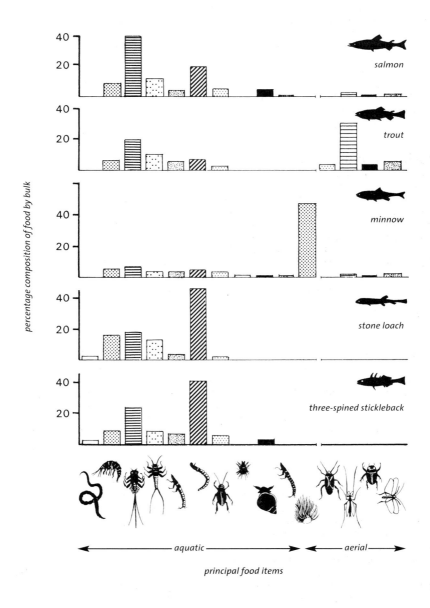

Fig. 21 Diet of five common fish in a stream community (River Endrick, Scotland)

Distribution

The distribution of fish species within a given geographic area can be considered at two levels: their general occurrence or absence throughout the area, and their detailed presence and abundance in specific waters there. The reasons for the general spatial distribution of any species are mainly historical, while specific sites where the species is successful are usually determined by ecological factors.

Although the status of a few of the species and subspecies mentioned in this book is in some doubt, approximately 316 species can be listed for Europe (c. 10 million km^2) as a whole. These are contained within 33 different families. This is a far less rich fish fauna than that of any other continent and the reasons for this are of considerable interest. For instance, North America (c. 24 million km^2) has 687 species in 34 families, and Africa (c. 30 million km^2) has 1,425 species in 50 families. Many of the reasons for this are found by a study of the history of the area, and in particular the impact of the Ice Ages.

During the last Ice Age, which came to an end approximately 10,000 years ago, almost the whole of the British Isles and much of northern and central Europe were completely covered by ice. This meant that at that time the fish fauna of Europe was made up partly of cold-tolerant forms which were able to survive in the lakes and rivers associated with the ice cap, partly of marine fish, and partly of fish living in southern Europe beyond the limit of the ice cap. It is likely that most of the indigenous species now making up Europe's fish fauna occurred in its southern waters at that time. In modern times there is very little permanent ice cover in Europe and many species of fish have moved into waters in the northern areas in particular. The European fish fauna as a whole can be considered as having originated from four different elements over the last 10,000 years: 1 Indigenous species which have gradually moved north from southern Europe. 2 Species with marine affinities which have colonised the area from the sea. 3 Species native to continents adjacent to Europe which have managed to invade European waters by natural means. 4 Foreign species (mainly from North America) which have been introduced by man. A similar story in reverse can be told for North America.

Indigenous species have moved about within Europe at different rates, and some fish obviously have much better powers of dispersal than others. This is an aspect of fish biology about which we know very little, and although various ideas have been put forward about the possibility of fish eggs being transported on the feet of waterfowl, by the action of waterspouts, of fish moving from one catchment to another via adjacent head waters, and of fish swimming from the mouth of one river to that of an adjacent one via stretches of fresh or brackish water along the coast, there is really very little firm evidence to support any of them – though all of them are likely to have taken place at one time or another. In general, the fish fauna of Europe becomes more impoverished as one moves from south to north or from the mainland on to adjacent islands, and it seems likely that an active, but very slow, process of natural dispersal is still taking place.

Many European fish are anadromous or catadromous, or at least able to tolerate fairly high salinities for considerable periods. It is quite easy to envisage these fish spreading rather rapidly as the ice cap disappeared, and indeed such fish are among the most widespread in Europe today, including trout, eel and sticklebacks. Some species which are purely freshwater at the southern end of their distribution have migratory races further north. The Arctic Charr is a good example of this, with many isolated freshwater populations in the British Isles and highland Europe, but anadromous stocks in Iceland, northern Norway, Greenland, Canada and the United States (Alaska).

Invasion of fish from adjacent continents is probably the least likely source of new species because Europe is separated from most other continents by considerable seas. The major land connection is to Asia in the east, north of the Caspian Sea. However, the high mountain ranges here, and the fact that northern Asia itself was similarly impoverished by the ice cap, has meant that there is little likelihood of fish entering in this way. The main movement of species has probably been in the Transcaucasian area, where a number of species occur which are found nowhere else in Europe.

Man has moved fish from place to place for many hundreds of years, but it is probably within the last 200 years that large numbers of significant introductions have been made in all parts of the world. Most introductions in Europe have come from North America, and of the list of species now established here includes several Salmonidae, Ictaluridae and Centrarchidae, as well as various other species which have been brought across the Atlantic. Changes like these, together with those caused by pollution and various fishery and land use practices, mean that most future distribution trends will be due to man's activities and not to natural agencies. This makes it essential for individual countries to consider the status of their native fish populations most carefully if their rarer endemic species are to be preserved. The fact that we are still very ignorant about the basic biology and distribution of many less common European species of freshwater fish means that considerable research remains to be done.

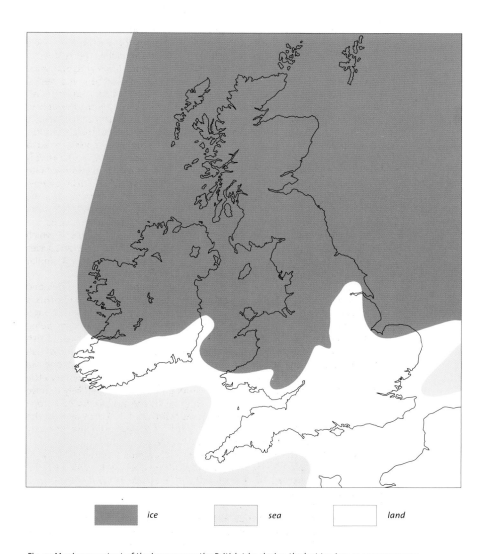

ice   sea   land

Fig. 22 Maximum extent of the ice cap over the British Isles during the last Ice Age, 10,000 years ago

Commercial fishing

It is very likely that man has eaten fish from early in his history, and evidence of this is available from prehistoric times in the form of cave paintings, stone etchings and certain types of stone traps. More recently, history has shown the importance of fish to man in abundant literary references, coats-of-arms, drawings, etc. Almost all sizes and types of fish are eaten, (except those which are actually distasteful or poisonous) ranging from several small species cooked and eaten whole as 'whitebait' to very large fish such as sturgeon and sharks. Fish may be eaten raw, or after preservation in various ways such as drying, smoking, salting, canning or freezing, although most fish is cooked before being eaten.

Methods

There are many different ways of catching fish and the methods used in fresh water are similar to those used in the sea, although usually on a smaller scale. Spears were one of the earliest methods of catching fish, and were very successful for flatfish like flounders in shallow water or large fish like Atlantic Salmon in small streams. Today spearing is carried out in Europe only as a form of sport – usually for flounders or for other species by scuba divers.

Baited hooks were also used very early in the history of fishing, originally being made of carved bone or thorns, but more recently mainly of metal. The size of the hook (like the size of the mesh in a net) is closely related to the size of the fish to be caught. In commercial fisheries long lines containing hundreds or even thousands of hooks, baited in various ways, but usually with pieces of fish or mollusc, are set for short periods (normally overnight) in suitable waters. Freshwater species caught in this way include European Eel, Wels Catfish, Burbot and, in the sea, Atlantic Salmon – though the method is of relatively little importance in commercial freshwater fisheries.

In addition to spears and hooks, simple traps were among early methods of catching fish and many, much more sophisticated types are in use today.

Fig. 23 Four types of commercial fishing net commonly used today

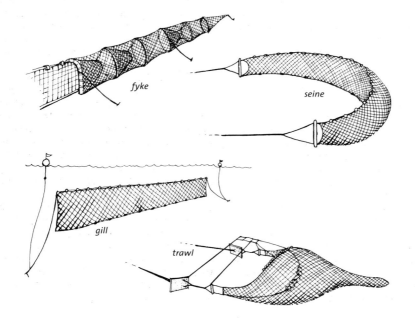

fyke

seine

gill

trawl

Almost all of these traps are based on some kind of funnel attached to an enclosure. The fish swim into the middle of the enclosure from which they have difficulty in swimming out. This is mainly because they swim round the outside of the enclosure and rarely near the centre where the small exit hole is sited. The earliest traps in Europe were made from stone or from wicker work. These were used to capture salmon, eels and other fish which tend to follow distinct routes (in which the traps were sited), especially during migrations. Most modern traps are made of netting, supported by stakes, weights and floats. Many of them are quite complex and include a long lead net which guides fish towards the entrance, and two funnel traps, one leading into the other. There is very little chance indeed of fish escaping from the innermost of these. Traps are still of considerable importance in many countries of Europe, such as Scotland and Ireland, for the capture of various fish species such as Atlantic Salmon and European Eel.

International statistics for commercial fish catches*

Geographic area	1975	1980	1985	1990
Freshwater fisheries				
Western Europe	286	373	426	465
Former USSR	948	792	905	975
World	6,673	7,183	10,169	13,717
Marine fisheries				
Northeast Atlantic	11,320	11,020	9,960	8,556
Mediterranean	1,050	1,500	1,720	1,150
World	58,790	63,520	75,890	83,942

*Figures given are in thousands of tonnes and include production by aquaculture.

The bulk of modern commercial fishing, however, is based on various types of fixed or moving nets (see fig. 23). Originally these were made from natural fibres such as wool and cotton, which were liable to rot, but most nets are now made from synthetic materials like nylon and terylene which are light, strong and much less likely to deteriorate. The manufacture of fishing nets is a skilled business and an important light industry in many countries such as Norway and Sweden. Three main types of net are in general use today in both freshwater and marine commercial fisheries. These are gill, seine and trawl nets.

Gill nets are sheets of netting hung in the water by means of an arrangement of floats, leaded lines and anchors. Since such nets depend on fish actually moving into them, they are normally made of very fine materials, often in a colour appropriate to the water in which they are being used. Many of them are most successful in poor visibility (i.e. in turbid waters) or in the dark. Gill nets depend on the fish swimming partly through an individual hole in the mesh and then becoming trapped by the meshes slipping behind the gill covers so that the fish can move neither backwards nor forwards. In other cases the fish may continue to swim forwards so that its body becomes securely wedged in the meshes. The size of the meshes used is critical in relation to the size of the fish caught.

Seine nets function on the principle of paying out a wall of netting from a boat so as to surround a certain area of water and enclose the fish therein; the method is a very popular and successful one. In shore seining, the net is set in a semicircle from a small boat and hauled to the shore by two groups of men. Shore seine nets may have a bag in the centre of the net to help to retain the catch. Most open water seines do have such a bag, and the net is set from a boat in a full circle before hauling. Two boats are often necessary in the operation. A rather specialised type of seine is the purse seine or ring net: this is cast in a full circle. It is normally much deeper than other seines and along the bottom of the net, running through large rings, is the purse line; as the net is hauled in, this line is tightened to close the bottom of the net.

Trawl nets are essentially open bags of netting hauled through the water at constant speed, normally from a powered boat. The main features of such nets are the shape and length of the bag, and the method used to hold the mouth of the net open. Two main types of trawl are commonly used in fresh waters: beam trawls, where one or more rigid poles keep the mouth of the net open; and otter trawls, where planing surfaces known as otter boards are used to keep the net open at either side while floats and a weighted line keep it open, top and bottom. The bags of such nets must be sufficiently long to ensure that any fish, having been overtaken by the mouth and moved back into the net, find it difficult to make their way out again.

Fish products

A high percentage of the body weight of fish (usually more than 50 per cent) is made up of muscle, or fish flesh as it is more commonly called. The most popular species for human consumption have a light-coloured flesh (usually white or pink) with few bones and a fairly mild taste or smell. Many less popular species are extremely bony or have powerful oily smells. The food value of fish flesh is high, comparing favourably with high quality beef or other meats. The main constituents of fish flesh are water, protein and fats, (very approximately in the ratio of 70:20:10) and a number of important vitamins. The fat content of many fish varies very much on a seasonal basis according to their condition, and the fats – unlike those of mammals – are unsaturated and therefore better suited for human consumption.

Since most European countries are adjacent to the sea, much of the fish eaten is of marine origin; it is really only in central Europe and one or two other areas that freshwater fish feature largely in man's diet. The number of species eaten in most countries is few, largely due to conservatism or prejudice against eating fish from local lakes and rivers. A few species are highly valued and may be classed as luxury foods. The most important among these is the Atlantic salmon.

Fish provide man with useful products other than flesh; oil is an important product from many fish – particularly Clupeidae and Gadidae – as are fish meals and fertilisers. These are obtained either from fish-processing factories where, after filleting, the waste parts of the fish are ground to make meal, or direct from fisheries where the fish are too small or of insufficient value to be sold for food. The meal is fed to poultry, pigs and cattle in combination with carbohydrates to give a balanced diet. The skin of some fish is used to make glue, while the swim-bladders of many freshwater species (for instance Atlantic Sturgeon and Common Carp) are an important source of isinglass. This is used mainly for clearing wines and beers. The scales of Bleak were formerly processed to produce an iridescent pigment used in the manufacture of artificial pearls.

Fish farming

The main aim of fish farming is the production, in large numbers, of fish for either the stocking of waters for sport or commercial fishing, or direct sale as food. In the former case, large numbers of small fish (sometimes even just eggs) are produced and sold for stocking. Increasingly, adult fish are reared, placed in natural waters and then caught almost immediately by anglers – a system known as 'put and take'. With table fish, the final objective is the size and quality of the individual fish produced for sale. The intensive nature of fish farming means that fish are kept under extremely unnatural conditions and fed on unusual substances. This has led to many problems in fish farming and combined with the fact that, although some types of fish farming are of very great antiquity, its science is very much behind that of most other agricultural sciences. This is because most meat comes from farmed sources, while most fish is still caught in the wild. Fish farming is rapidly becoming a very important industry, however, and significant advances are being made every year in such fields as genetics, disease and feeding.

The Chinese and the Egyptians reared fish many hundreds of years ago, and the Romans also regarded this as an important aspect of agriculture. In Europe in the Middle Ages most fish farming was carried out by religious orders in monasteries, Common Carp being the most important single species. Within the last hundred years, and particularly in the last ten years, fish farming in Europe has increased greatly, as has the number of different species which are now farmed. There are a number of reasons for this expansion: the shortage of food, especially protein, which has meant that more land and water must be used for its production; the availability of fertilisers and new types of food for improving growth; and the increasing demand for luxury foods in many countries. The recent increase in salmonid production for the table is a good example of this.

A wide variety of freshwater species are now reared in Europe. Many of the fish involved are not native species, but have been imported from North America and Asia especially for rearing. There is no doubt that further introductions will be made, and that fish genetics will play an increasing part

Fig. 24 A typical modern floating cage fish farm. Access is along the central deck and fish are fed by automatic feeders

in fish husbandry, just as it has with other domestic animals. Not only will strains of particular species be selected which show especially desirable qualities of growth, resistance to disease, etc., but hybrids of many species will be produced in an attempt to produce fish ideally suited to intensive farming.

Most fish farming is carried out in artificial ponds, and often in cages in ponds. Sometimes natural waters (usually small lakes) are used, and cages or large enclosures are increasingly placed in lakes and in the sea. Ponds, especially those specifically designed for fish farming, have overwhelming advantages over other types of water. They can be emptied or filled easily. fish can be caught without difficulty by netting or emptying, and feeding is simple. Predators such as herons and gulls, and diseases (including parasites), are much easier to control. Ponds used in fish farming can vary in size from a fraction of a hectare (1 hectare is equivalent to 2.47 acres) to hundreds of hectares; the majority in Europe are between 0.1 and 1 hectare in surface area.

Two major points must be remembered when a fish farm is being planned: the ground and the water. Economically, sandy, marshy or other poor quality (agriculturally speaking) ground is often best, but not always appropriate. Soil which is too porous is often unsuitable, and the best sites lie on ground which is impermeable, or which is permanently waterlogged (but not subject to flooding). The slope of the ground is also important. Ideally, the ground should be sufficiently low at some point – preferably near a river or some other watercourse for the ponds to be able to be drained without pumping. Ideally either the water supply available should also have pressure, or the source of water to the fish farm should be sufficiently high above it to provide a reasonable flow.

The water used for the farm should be free from pollution and adequate in quantity and quality for the type of fish farming being carried out. Broadly speaking, the fish species being reared fall into two ecological categories: salmonids and related forms which require large amounts of cool (lower than 20°C or 68°F), pure water with a high oxygen content, and cyprinids and similar fish which are much less exacting in the amount and quality of water, including oxygen, needed. They also require far higher temperatures (preferably higher than 20°C) for much of the year if adequate growth is to be attained. There is a tendency for salmonid culture to predominate in fish farms in the highland and northern areas of Europe, and cyprinid culture to be of greatest importance in lowland and southern Europe.

As well as matching the environment of the fish farm to that required by the species reared, the farming procedure must also be geared to the biology of the fish. The rearing process is normally carried out in three stages: incubation, rearing of fry, and rearing of juveniles to adults. Although the eggs used on fish farms are sometimes obtained by natural spawning in ponds, or even in the wild, most species are captured at spawning time and stripped of their eggs and sperm. These are mixed together in a container and left for a few minutes before being washed and placed in containers for incubation.

Most incubation takes place indoors where precise conditions of water flow and quality (especially temperature, oxygen and silt content) can be maintained. Eggs are placed, usually in thousands or even millions, in shallow troughs or open jars in which a flow of water is maintained throughout incubation. In addition to maintaining appropriate conditions for the species being reared, each batch of eggs must be examined regularly to check that development is taking place and to remove dead eggs which can rapidly rot and kill off other members of the batch. After hatching, which may take from 5–10 days in carp but up to 100–150 days or longer in trout and salmon, the larvae are normally removed to another part of the farm.

The larvae may not require feeding for several days if the yolk sac is large after hatching, as is usually the case with salmonids. As soon as feeding commences, however, adequate quantities of food must be supplied regularly, usually several times a day. This must be adequate for the stimulation of good growth in the larvae without causing deoxygenation. Old decaying food combined with fish faeces will quickly rob water of its dissolved oxygen, leading to high fish mortalities. Adequate supplies of water which bring in oxygen and carry off wastes are most important at this stage. Having changed from larvae to fry, which may vary in size from 3–10 cm (1–4 in) or more, fish are removed for the next stage of the rearing process.

It is at this point that most fish used for stocking are taken and liberated in natural waters. This is usually carried out by lorries loaded with special tanks and oxygen supplies, but it is not uncommon for remote waters to be stocked from aeroplanes or helicopters. Many fish-stocking programmes are correctly criticised by fishery biologists, and considerable research is still required on this subject. It is known, for instance, that for many days after being released, hatchery fish find difficulty in adapting to natural surroundings and are much more vulnerable to predators and less able to find food than wild stock. In contrast, once such fish become established in sufficient numbers, they may compete for food and space with native stocks.

The final phase in fish farming is that of rearing the fry until they are large enough to be sold for the table, or are of a catchable size for selling to angling syndicates for 'put and take' fisheries. This stage is the most expensive of the rearing process for it requires the greatest input of the three most valuable commodities – space, food and manpower. Almost all species require to be fed regularly, usually two or more times a day. The food varies according to species but is often very expensive – particularly for salmonids, which require a protein-rich diet. This food, usually in pellet form, is often based on fish meal obtained either as waste from fish-filleting plants or directly from species harvested for the purpose. During the final rearing stage the fish are regularly graded for size so that only those of similar size are kept together. A constant watch is kept for parasites and disease, and many ponds are cleaned or drained regularly to keep these troubles to a minimum.

Fish farming in Europe is still expanding, but much of it will not answer the world's food shortage, for the systems concerned are based on high-quality protein diets, much of which is wasted in the conversion to fish flesh. These systems, in times of shortage, can only lead to luxury foods and can never be considered as a source of cheap protein. The few herbivorous fish at present being reared in Europe can convert vegetable (often waste) material to high-quality protein, and it is these systems which should be encouraged if the principal search is for cheap food.

The principal species of freshwater fish cultured in Europe

Common name	Scientific name	Purpose of culture
Sterlet	*Acipenser ruthenus*	food
Atlantic Sturgeon	*Acipenser sturio*	conservation
Allis Shad	*Alosa alosa*	food
Chum Salmon	*Oncorhynchus keta*	food/stocking
Pink Salmon	*Oncorhynchus gorbuscha*	food/stocking
Atlantic Salmon	*Salmo salar*	food/sport
Brown Trout	*Salmo trutta*	food/sport/stocking
Rainbow Trout	*Oncorhynchus mykiss*	food/sport/stocking
Arctic Charr	*Salvelinus alpinus*	food
American Brook Charr	*Salvelinus fontinalis*	sport/stocking
American Lake Charr	*Salvelinus namaycush*	sport
Vendace	*Coregonus albula*	food
Common Whitefish	*Coregonus lavaretus*	food
European Grayling	*Thymallus thymallus*	sport
Northern Pike	*Esox lucius*	food/sport
Roach	*Rutilus rutilus*	sport
Belica	*Leucaspius delineatus*	bait species
Orfe	*Leuciscus idus*	ornamental pond fish
Minnow	*Phoxinus phoxinus*	bait species
Rudd	*Scardinius erythrophthalmus*	sport
Tench	*Tinca tinca*	sport
Bleak	*Alburnus alburnus*	bait species/artificial pearls
Crucian Carp	*Carassius carassius*	food
Goldfish	*Carassius auratus*	ornamental pond fish
Common Carp	*Cyprinus carpio*	food/sport
Grass Carp	*Ctenopharyngodon idella*	farming/weed control
Silver Carp	*Hypophthalmichthys molitrix*	farming/algal control
Bighead Carp	*Hypophthalmichthys nobilis*	farming
Wels Catfish	*Silurus glanis*	food/sport
European Eel	*Anguilla anguilla*	food
Striped Grey Mullet	*Mugil cephalus*	food
Thinlipped Grey Mullet	*Liza ramada*	food
European Perch	*Perca fluviatilis*	food/sport
Pikeperch	*Sander lucioperca*	food/sport

Sport fishing

No outdoor sport has more active participants than angling. The most recent survey indicated that more than 4 million people went fishing at least once a year in Britain alone, and France is said to have 5 million anglers. A huge industry has developed to meet their needs, with much of the tackle being produced in Britain, France and Sweden, and most of the nylon monofilament line being manufactured in Germany. So vast is the business associated in one way or another with angling that if the sport suddenly ceased it would affect the livelihoods of a great many people. Such a disaster would not only affect fishing tackle manufacturers, but all levels of the fishing industry from the breeder of maggot bait to the owner of a top salmon beat who charges £1,000 or more for a week's sport.

The sport of freshwater angling is divided into two basic kinds – coarse fishing and game fishing – and both kinds subdivide into various separate interests. Within coarse fishing there are match anglers who fish against each other purely for the competitive element, or for money or both, so-called 'pleasure anglers' who fish for anything they can catch, and specimen hunters who concentrate upon the capture of exceptionally big fish. Sometimes they will even concentrate on one single species, and there are clubs for specialists in the pursuit of Pike, Carp, Tench and Eel.

Game fishing is rather less fragmented, but there are those who specialise in migratory fish such as Sea Trout and Atlantic Salmon, and others who find their paradise in river fishing for Brown and Rainbow Trout. The most recent developments in game fishing are trout fishing in reservoirs, and small-water fishing for very big fish – Rainbow Trout reared to record breaking proportions

Fig. 25 Coarse fisherman netting his catch from a lowland water

or with special colour forms ('golden trout' or 'blue trout') by fishery and fish farm owners with an eye for publicity.

Where relevant and available, 'Rod Caught Records' have been included in the species accounts which follow in the main body of this book.

A comprehensive description of the tackle that anglers need, and the many methods employed by them, would fill a large book and numerous volumes are available on the subject. The most universal item of tackle is the rod, which is now made almost exclusively of fibre glass, although carbon fibre has also made an impact, especially in reservoir angling. Formerly rods were made from various types of wood, and the best rod in that material was hollow-built split bamboo. Every rod is capable of handling a limited range of line. Line which is too weak for a rod will break on striking a fish, and there is little point in using line strong enough to bend a rod double. Casting with fly tackle is a precision business with the objective of presenting a recognisable imitation of natural food as near as possible to the nose of a feeding fish or to provoke a fish into snapping at an offering which intrudes into its domain.

The angler who fly fishes, or who seeks his quarry with artificial lures, or live or dead fish, can travel very much lighter than the angler who fishes with bait. The bait fisherman must carry a holdall filled with rods, an umbrella, landing net handle and a variety of metal poles for supporting the rod, and a basket which contains boxes of floats, lead weights, reels, groundbait, hooks bait and many other accessories, too numerous to mention. The basket doubles as a seat, and the canvas bag he also carries usually transports a keepnet for retaining his catch, and the head of the landing net to scoop the bigger fish out of the water. The angler's bait is also varied – ranging from the traditional large tin of maggots or casters (pupae of the maggot) and worms to the high protein boiled baits used by carp fishermen.

It will be seen from what has been said about sport fishing that it can be an involved and exciting sport, depending not only on the angler's tackle and how it is used, but on weather, the water and the mood of the fish. It has the capacity to exercise the mind of both adults and children. It is because pleasure can be derived from angling at so many levels that the sport is so popular. It is claimed, perhaps with some truth, that man goes fishing because he has not entirely forgotten the hunting instincts his ancestors once needed in order to survive. If this is so then his motive has certainly changed, for very few anglers actually fish for food. Coarse fishermen rarely eat what they catch. They retain their fish in nets and release them at the end of the day or after the match. Game fishermen certainly do eat at least some of what they catch, but it cannot be said that they fish simply for food, for this would result in very expensive meals due to the high cost of a day's trout fishing.

Man is still a hunter, and in those where the hunting instinct remains deeply buried, there is still a spark of interest in what the hunter catches. Witness the crowds which gather on the harbour wall when the fishing boats come home. But the anglers' hunting instinct is now converted into a sport, and because of man's penchant for organising almost anything, it is inevitable that angling is organised. Thus there are small clubs, large clubs, national associations and federations, and in turn there are links between the national bodies of various countries and their counterparts in other parts of Europe, North America and the rest of the world.

Aquaria and ponds

To many people the word 'aquarium' conjures up little more than the idea of a sterile goldfish bowl and its prisoner, or a brightly lit tank in a restaurant with multi-coloured plastic gravel, plastic plants and sometimes even plastic fish! Aquaria mean much more than this, however, and to many thousands of people in Europe and elsewhere form the basis of an absorbing hobby. In other spheres aquaria offer a valuable educational tool, or the basis for exciting work on processes ranging from animal behaviour to research on cancers. With modern technology it is possible to keep almost all species of fish alive in aquaria, with the exception perhaps of those which live in very deep water.

Aquaria are employed in two main ways: firstly, simply as containers for keeping alive fish which are required for some purpose such as food or experimentation. Secondly, as small isolated habitats replicating those found in nature and including not only fish, but also aquatic plants and invertebrates. Such aquaria can be both educational and aesthetically pleasing, as can many types of outdoor pond (see fig. 26).

The object of experimental aquaria is to test the performance of fish under various, often extreme conditions. Thus light, temperature, water flow and quality (including oxygen content), are all variables which may be altered during experimentation. Some of the work in this field is of a very theoretical nature, but much of it is totally applied and relevant to man's needs. The study of fish genetics and the value of various types of foodstuffs are two examples of research related to priorities in fish farming at the moment. One of the obvious impacts of pollution on fresh waters is the death of fish – often in very large numbers. Indeed, were it not for the death of fish many people would not know (or even care) that a stream was suffering from severe toxic pollution. Attempts are being made at the moment with various species of fish in aquaria to develop a 'standard fish', whose precise reactions to various polluting substances, and to combinations of these, are known. This allows the possibility of biological assay of effluents, etc. by introducing them to fish in aquaria, or suspending fish in cages in the effluent channel. Such tests have a number of advantages over existing, purely chemical methods of detecting pollution, mainly in the ways in which they act as

Fig. 26a A small concrete garden pond dominated by water lillies

Fig. 26b A large landscaped garden pond with an attractive shoal of bright Koi Carp

integrators both of time, and of several different pollutants which may be present. Rainbow Trout is the species most in use for this purpose at the moment.

The main attraction of the indoor aquarium is that it represents an easily observed microcosm, with conditions similar to those found in nature. The object of keeping such aquaria is to achieve a balance between the various components, so that something equivalent to natural recycling can be attained. It is rare for this to happen completely, and normally some material is added to the system in the form of fish food and some removed in the form of fish droppings and decaying or excessively growing plants. Nevertheless something near a balance can be achieved.

The plants in an aquarium are normally rooted in gravel and are able to use up quantities of waste salts and carbon dioxide produced by the fish. Under the action of light the plants photosynthesise and produce oxygen (much of this also comes through the surface of the water) which is respired by the fish. Certain fish eat parts of the plants and also algae, which grows on all surfaces in the aquarium, and the ecosystem is often made more complex by adding invertebrates such as molluscs and crustaceans, which may help to prevent excessive plant growth and whose eggs and young form a source of food for the fish. Many very useful books have been written about aquaria, and some of these are listed in the bibliography. In addition there are many hundreds of aquarium societies all over Europe and anyone with an interest in this subject is recommended to join one.

The educational value of fish in aquaria is something which should not be underestimated, for if appropriate species are kept in the right conditions they may be persuaded to go through their full life cycle in one year and to demonstrate a wide range of biological principles which are relevant to the theory being taught in the classroom. This may include many aspects of behaviour, reproduction, ecology, anatomy and physiology. Common European species which are of great value as educational aids since they may be kept and bred in aquaria include the Mudminnow, Minnow, Bitterling, Goldfish, Three-spined Stickleback, Mosquito Fish, Rock Bass, Pumpkinseed, Chanchito and Bullhead.

The value of aquaria in teaching is well known, and they are a common feature of many educational establishments. Their value to the general public is less appreciated and it is a lamentable fact that most inhabitants of European countries have seen only a few of their native fish, and these only too often on a fishmonger's slab. Fish are among the most difficult of animals to observe in nature and can only be fully appreciated when observed in aquaria of sufficient size, set up according to their ecological requirements. The few really good public aquaria in the world today (for instance in Amsterdam and Vancouver) show that it is quite possible to display the range of native species found in a country and that these are often more attractive and exciting than the standard brightly coloured exotic species so commonly featured in other public aquaria.

Collecting
and preserving

Depending on the species concerned, fish may be caught by a wide variety of methods. Among the commonest methods of capture are angling for large fish, and the use of a hand net for smaller fish. There are many other methods of capture (some of them much more efficient, but illegal in most situations) which can be used. These include electrofishers, poisons, gill nets, seine nets, trawl nets, and traps, which are described in detail elsewhere. Most methods of fishing are highly selective and may often capture only one size group of one species, and sometimes even only one sex. In carrying out any detailed study of a mixed fish population in a habitat, it is normally advisable to use several different methods of capture.

As noted elsewhere, the major key in this book will not serve to identify the very small specimens of most species. For a short time after hatching the young of most fish are changing very rapidly in form, and features characteristic of the species do not appear for some time. Unlike many invertebrates, adult fish, because of their large size, can be identified in the field, and so there is no need to take them to the laboratory. Thus, it is possible to identify specimens correctly immediately after capture and then return them alive to the water. There is little point in killing such fish unless they are required for food, research or some other legitimate purpose. In the case of species which are difficult to identify accurately in the field it may be necessary to take them away for detailed examination. It is preferable that they be kept alive for this purpose, but this is often not possible, especially with large or delicate specimens, and most of these fish have to be killed. Clearly all specimens must be killed where dissection is essential for their accurate identification.

Killing
One of the best methods of killing fish – humanly and without damaging them – is to use a liquid anaesthetic. Specimens dropped into a suitable solution are narcotised very quickly and can then either be frozen or transferred to a suitable fixative. Ideally, fish should be examined as fresh as possible – especially where colour is an important feature.

Freezing and fixing
Frozen fish retain their colours much better than fixed specimens; the best procedure for freezing is to place each specimen in a polythene bag with a little water and a suitable label and freeze this in its entirety as quickly as possible. The specimen should be kept straight during this process and care must be taken not to damage its fins.

Even frozen fish will not keep indefinitely, however, especially if they have to be subjected to periodic thawing for examination purposes, and normally specimens to be stored must be fixed in some way. The most useful fixative is 4 per cent formaldehyde. Each fish should be preserved by placing it flat on its side in a shallow dish with its fins spread as much as possible, and then pouring enough of this solution over it to cover it completely Specimens should be left for several days to ensure complete fixation. With fish more than 30 cm (1 ft) in length, it is normally advisable to make a small slit in the ventral body wall, or to inject the body cavity with a small amount of 40 per cent formaldehyde. This ensures complete fixation internally, after which the fish can be stored temporarily in polythene bags or permanently in suitable jars, either in 4 per cent formaldehyde or in less offensive preservatives such as 70 per cent alcohol or 1 per cent propylene phenoxetol. Each jar should have inside it a label, written in pencil or indelible ink, with a note of the species concerned, the water where it was collected, the date and the name of the collector.

In situations where the suggested preservatives are not available, quite good results can be obtained by using emergency materials which are normally easily obtained. For example, reasonable preservation can be obtained with: 1 Methylated Spirits, diluted 7 parts of spirits to 3 parts of water; 2 Salt, diluted 1 part of salt to 2 parts of water; 3 Vinegar (acetic acid), as used as a condiment. Preservation can be carried out satisfactorily in polythene bags, again keeping the fish straight.

Eggs or larvae can be preserved and stored in small tubes containing either 4 per cent formaldehyde or 70 per cent alcohol. Labels with relevant data (species, locality, date, colour of eggs when fresh, exact habitat, name of collector, etc.) should be placed inside each tube. In the case of eggs, a reasonable number should be taken where possible, especially if the eggs are adhering to each other, because the form of attachment may be important in identification.

Some species of fish can be identified from their scales alone, and in addition it is often possible with several good scales taken from the side of the body to establish the specimen's age, and certain features of its history. When scales are available from specimens they should be placed inside a small envelope which is flattened and then allowed to dry; they will keep indefinitely in this way. The envelope should bear on the outside the following data: species, locality, date, name of collector, length, weight and sex of the specimen concerned. For identification, the scales should be cleaned and mounted on glass slides, either dry or in glycerine jelly. The length of a fish can be read in a number of different ways; the most useful method is to record fork length – the exact distance between the tip of the snout and the tip of the middle ray of the tail fin (fig. 18).

As with other vertebrates, bones are often very useful in the identification of fish, and in some cases (e.g. with the pharyngeal bones of Cyprinidae and the vomer bones of Salmonidae) they may be essential for accurate identification. Certain bones are also of major importance in providing information on the age and growth of some species (e.g. the opercular bones of Percidae and Esocidae). Preparation of all such bones for examination and subsequent preservation is a relatively simple task; fresh or frozen material should be used – not material which has previously been fixed. The relevant bones should be dissected out from the fish, along with their attached tissues. Each bone should then be dropped into very hot water for a few minutes and then scrubbed gently with a small stiff brush to clean away soft tissue. The process should be repeated until the bone is completely clean; it can then be placed on clean paper and allowed to dry out slowly in a warm atmosphere. Details concerning the fish from which the bone was removed (species, locality, date, name of collector, length, weight and sex), should be written either on a small stiff label attached to the bone by strong thread, or on the outside of an envelope or small box in which the bone is kept. Bones cleaned and dried in this manner will keep more or less indefinitely.

Although it is possible with the aid of this key to identify down to species all fish known to occur in fresh water in Europe, a few species are rather difficult to identify accurately without experience, and there is also the possibility of the occurrence of a hybrid, or of a species new to a country. In cases of doubt one or more specimens of the species concerned should be killed and preserved as described above, and then sent together with relevant details to a competent ichthyologist for examination. It is not normally advisable to send fresh specimens by post, as they deteriorate too rapidly. When sending preserved specimens they should be drained of preservative, wrapped in damp muslin and then sealed inside a polythene bag. If this is finally placed in a box with packing, and then parcelled, the fish will travel for several days in perfect condition. Usually, the national museum of each country is the best place to lodge rare specimens or those which are difficult to identify.

Threats to fish

Fish populations have been utilised by man for many thousands of years, and it is often difficult to separate the effect of his impact from changes due to more natural processes. Over the last two hundred years, particularly the last few decades, many new and intense pressures have been applied, and in most cases these have been detrimental to fish conservation as a whole. Inevitably, many of the threats are interlinked, the final combination often resulting in a complex and unpredictable situation.

Fisheries

The impact of fisheries (both commercial and sport) on the populations which they exploit can range from total extinction of the populations to the more or less stable relationship of recruitment, growth and cropping which exists in many old-established fisheries. The essence of success in fisheries is to have a well-regulated fishery where statistics on the catch are monitored continuously and used as a basis for future management of the stock. It is clearly in their own interests for those concerned with fisheries to adopt sound policies in relation to the close season, the type of gear used, and the numbers and sizes of fish caught, etc., so that the stocks will survive. Although there are considerable gaps in our knowledge of the population dynamics of many fish species, we have sufficient information to suppose that well-regulated fisheries, whose management is based on existing scientific information, should prove sustainable for the species of fish concerned. Yet all too often freshwater fishery management is still not based on sound ecological information and is consequently unsustainable.

Stocking was once considered by many (and is still thought by some) to be the main management tool to be used in fisheries; a large number of derelict fish hatcheries and ponds in various countries provide evidence of the faith of past generations in this procedure. There is no doubt that stocking is a valuable part of the management policy of some fisheries, but it is only necessary in waters where spawning and nursery areas are absent or inadequate to provide recruitment for the losses due to the angling pressure involved. The numbers of waters where this is true are increasing, and are likely to continue to do so, resulting in the creation of more and more 'put and take' fisheries. Fish food production in these is largely irrelevant, as the fish introduced are of a catchable size and likely to be caught before they starve. There are very few aquatic ecosystems into which a species of fish could be introduced and established without some major alteration within the system itself. This being the case, all proposed introductions of fish, either as species that do not occur in Europe or that would be new to a water or catchment system within it, should be given the most careful consideration.

With native species of restricted distribution, or foreign species already well established in Europe, careful consideration should be given to any proposals to introduce them outside their present areas of occurrence. Many native species appear to be gradually distributing themselves by natural means, for example European Perch and Northern Pike, but even in these instances the effect may be drastic. Thus several cases are known of Northern Pike gaining access to waters containing populations of Brown Trout, and completely eliminating them. Many species, too, have been dispersed by man. The introduction of Dace and Roach to Ireland, and their influence on other fish there, have been carefully documented, while recently the Common Barbel has been moved well beyond its former area of occurrence in Great Britain. Many hundreds of introductions take place annually, mostly within the existing distribution boundaries of the species involved.

In the case of fish species that are new to Europe, there are several points which should be taken into account before introduction is contemplated: 1 It should be quite clear that there is some real purpose behind the introduction, such as sport or aquaculture. 2 It is undesirable to introduce species which are likely to have a major influence on ecosystems through predation, competition, or feeding habits (such as destruction of vegetation). 3 Consideration should be given as to whether or not the species could be controlled readily. Some species may not breed because of climatic conditions, while others may have very specialised requirements for spawning, etc.

There are without doubt some circumstances in which translocation of a native species may be justifiable. Obvious cases include the introduction into clean waters of fish whose former populations have been destroyed by pollution of some kind, and the establishment of suitable species in newly created waters such as reservoirs. Some natural waters contain no fish species, and while it is scientifically desirable that some of these remain like this, a case could be made for the introduction of certain suitable fish into others. It is particularly undesirable, on the other hand, that any new species be introduced to waters containing rare native species. Rare native species may well be moved to other waters as a conservation measure, however.

Pollution

Pollution of fresh waters is a subject which is widely discussed and will not be dealt with in detail here. Suffice to say that most pollution comes from domestic, agricultural, or industrial wastes and that it can be either toxic, thereby eliminating all the fish species present, or selective, killing off only a few sensitive species or altering the environment so that some species are favoured and others are not. Eutrophication (see below) is sometimes thought of as mild pollution.

Because of the large numbers of rivers and proximity to the sea, there is relatively little direct pollution of standing waters compared to running waters in Europe. Most effluents are directed into running waters and estuaries (or the sea), and it is the species concerned with such systems which are mostly influenced by pollution. Species with marine affinities are especially affected, for it is normally the lowest reaches of rivers and their estuaries that are most seriously polluted, but through which such fish must pass at one stage in their life history. Thus one extreme belt of pollution in a river system can affect the fish population of that whole system by acting as a barrier to these fish.

The process of eutrophication in a wide array of waters has received considerable attention in many parts of the world. Essentially, the phenomenon is due to increasing amounts of nutrients entering the water from geological, domestic, industrial and agricultural processes. The general changes which take place in a body of water that is undergoing eutrophication will not be discussed here, except to mention that they involve large increases in the abundance of algae (associated with this is a change in the species composition, and subsequent changes in the benthos and fish populations).

As far as fish are concerned, one common response to increased eutrophication is a marked increase in growth rate. There appears also to be a direct relationship between accelerated eutrophication and an increased rate of parasitism in fish. With the change from clean oxygenated substrates to silted, sometimes poorly oxygenated, substrates in waters undergoing eutrophication, many fish lose valuable areas of spawning ground. This has a serious effect on species which spawn in deep water, such as whitefish. In lakes that are undergoing normal eutrophication in nature, there appears to be a succession of species and dominance from salmonids (for instance charr) to coregonids (for instance whitefish) to percids (for instance perch) to cyprinids (for instance carp). This succession, usually considered undesirable from a game angling point of view, is considerably accelerated by artificial eutrophication. The mechanisms which bring about these changes are still rather poorly understood.

Acidification

Acid precipitation has caused major damage to fish stocks in various parts of the world including North America, Scandinavia and the British Isles. Salmonid fish are particularly vulnerable, but many other fish species have been affected. A survey of fish populations in 50 lakes in southwest Sweden showed that, in acidified lakes, Brown Trout, Arctic Charr, Roach and Minnow were all affected. Atlantic Salmon have been completely eliminated from several Swedish rivers over recent decades. Acid precipitation has also devastated fish populations in southern Norway where the principal fish species affected are Atlantic Salmon, Brown Trout, Arctic Charr and Brook Charr.

One of the most characteristic effects of acidification on fish populations is the failure of recruitment of new age classes into the population. This is manifest in an altered age structure and reduction in population size, with decreased intra-specific competition for food and increased growth or condition of survivors. To anglers, the fish stock then appears to 'improve' because larger fish are caught, albeit in lower numbers. However, with no recruitment, year by year, the population contains fewer fish until eventually there are none.

Afforestation

The impact of coniferous afforestation and forestry practice on freshwater habitats in Europe and North America has caused much concern in recent years. The effects of each stage of the forestry cycle – ground preparation, tree planting to canopy closure, the maturing crop and felling – may have an impact on local fresh waters. The physical aspects of afforestation affect: (i) the hydrology of streams, as shown by (a) increased loss of water through interception and evaporation from the coniferous forest canopy, and (b) a tendency to higher flood peaks and lower water levels during droughts; (ii) the release of sediments to streams because of erosion following ploughing and weathering of exposed soils; (iii) reduced summer water temperatures in afforested streams where the channel is shaded.

The principal chemical changes in fresh waters in afforested catchments include: (i) increased nutrient levels from the leaching of exposed soils and applied fertilisers, and (ii) the acidifying effect of air pollutants (especially sulphates) which are intercepted by conifers (as airborne particles) and then transferred to the ground (by rain – itself often very acid) and eventually to adjacent water courses. The base-richness of local soils and rocks is of major importance here, for it is only in areas lacking basic ions that acidification occurs. Much of the biological damage is due to the high amounts of aluminium (leached from the acid soils).

These physical and chemical effects combine in various ways to affect the plants and animals of fresh waters in afforested areas. Changes in hydrology and ambient water temperatures tend to make conditions in streams more extreme for most biota. Turbidity decreases plant growth through reduced light penetration and physical siltation. Increased nutrients alter plant communities and cause problem crops of algae in streams and lakes. Acidification of water courses affects the composition of their plant and invertebrate communities and completely eliminates fish in some cases. Other vertebrates, such as amphibians and birds, may also disappear.

Engineering schemes

It is well known that hydro-electric schemes can have deleterious effects on local fisheries and this is certainly the case with Atlantic Salmon in some waters. Arctic Charr seem to be less affected than other salmonids, and there is evidence that some stocks (which are mainly plankton feeders) may be favoured by the fluctuating water levels, which adversely affect Brown Trout. It is believed that, because the fluctuating levels often devastate the littoral flora and fauna, the trout population, which mainly feeds in the littoral area, is also adversely affected. The plankton, on the other hand, is not affected and so Arctic Charr still have their main food source.

Water supply schemes may also have significant effects on fish populations, especially where large volumes of water are transferred from one catchment to another. In England, for instance, large numbers of Arctic Charr and Common Whitefish are pumped out of Haweswater each year as part of the water system supplying Manchester. The impact can be particularly serious where no account is taken of local fish ecology when the engineering works are being designed. This can lead to serious damage to fish stocks as, for example, at Loch Lee in Scotland, where substantial numbers of adult Arctic Charr are washed out of the loch each year at spawning time, due to construction of a spillway near charr spawning grounds.

Land use

The impact of different forms of land use can be important to fish populations in various parts of Europe. The creation of reservoirs for hydroelectricity or water supply is an obvious example, where shallow streams or small ponds are transformed to large, often deep, reservoirs. In some cases these reservoirs have highly fluctuating water-levels, and this can have a significant influence on littoral populations, including fish, which either live or spawn in shallow water or inundated tributaries. The great majority of standing waters in Spain and some other European countries are artificial and the dams of such reservoirs normally act as barriers to migratory fish. Any remedial measures in such situations are directed towards the conservation of anadromous species of sporting value and in Scotland, for example, such measures include a guarantee of minimum river flow, the provision of freshets, construction of fish passes, erection of smolt screens, creation of artificial spawning gravels, use of hatcheries, and the opening up of previously unused natural spawning gravels. Although some impoundments may improve the quality of sport fishing, practically never is thought given to species other than those of sporting or commercial value.

One serious problem related to land use is the widespread destruction by draining or filling of many thousands of small ponds, ox-bow lakes, and other minor waters in all parts of Europe – particularly in lowland areas. Such habitats are of major importance to populations of smaller fish species (as well as being important aquatic communities in their own right), but rarely is there any outcry about their destruction. The economic and social claims of those destroying them are well known (cheap dumping grounds for garbage, reclamation of land for agriculture or industry, and removal of a potential 'nuisance' and danger to children), and apparently outweigh all else in the minds of those concerned. In a few areas the trend is the reverse, where new waters are dug out for water shortage, for agriculture or the winning of gravel. Although the original aim is an economic one, such waters can be important in conservation terms if constructed and managed intelligently, and they are becoming an increasingly important amenity in some areas.

Conservation

The maps included in this book show the present known distribution of each species of freshwater fish established in Europe. Several species are increasing their range of distribution, whereas others are decreasing in this respect. A few are in danger of extinction – mainly as a result of stresses imposed by man. There is a considerable amount of fundamental research still required, especially on the less common species, if all interests in the resource are to be protected and the extinction of rare species is to be prevented.

The conservation of freshwater fish means a number of quite separate things to different people. Most of the discussion on fish conservation in the past has centred around species which are of direct importance to man either commercially or for sporting purposes. The number of people involved in these activities is very large (about 4 million in the British Isles alone), and forms a significant proportion of the population.

In considering the conservation of fish populations, there are a number of other important aspects which have received far less attention – presumably because they appear to have no economic importance or, if it is admitted that they have such importance, because it is too difficult to assess in terms of 'cash value'. Amenity is one such reason. The amenity value of water bodies is an accepted fact now, and one of the attractions of being near them is to see fish swimming and leaping in clear water. For many people it is psychologically satisfying to know (without necessarily seeing) that a water body contains healthy, stable fish populations. As already mentioned, fish are excellent indicators of the degree of pollution of a water body: fish deaths are often one of the first obvious indications of sudden pollution, and waters which are fishless owing to pollution are generally considered to be unsatisfactory – regardless of their visual condition. Usually among the top consumers in food-chains in aquatic systems, fish often provide a very sensitive measure of pollution of the environment by insecticides and other insidious poisons which build up in the food chain and accumulate in their tissues.

The recreational value of fish populations is often equated solely with their use for sport fishing, whereas this is often far from the case. The economic element and adult participation in sport fishing often mask the fact that large numbers of children (and their parents!) gain very great recreational pleasure from catching small fish and later keeping them in ponds or aquaria. The range of important species in this context is far wider than that which is appropriate to sport fishing. Related to this activity is the educational value of these species to schools and universities – especially from behavioural and anatomical viewpoints. The ecological relationships within waters can also be important as far as fish are concerned. Thus, for example, they often form the main food supply for interesting or rare predatory birds.

In a scientific context, fish populations have a number of uses, and are widely employed in a variety of research studies. One important aspect in this field is the maintenance of populations of rare species or isolated populations of common species which may possess unique gene pools.

Active conservation

The existing protection given to most native fish in Europe is inadequate in terms of management, legislation and the establishment of appropriate reserves. The exceptions mostly relate to fish of angling importance which are given substantial protection both in the water and through available legislation. Even here, however, the situation is not entirely satisfactory for virtually no reserves have been established primarily for native fish conservation and in most countries there are only a few pieces of appropriate legislation.

There is an enormous amount of work to be done in the field of fish conservation. In addition to establishing the status of fish in each geographic area, much effort must go towards identifying the conservation needs of the most endangered species and implementing these as soon as possible. As well as habitat restoration, one of the most positive areas of management lies in the establishment of new populations, either to replace those which have become extinct or to provide an additional safeguard to those still extant. Any species which is found in only a few waters is believed to be in potential danger and the creation of additional independent stocks is an urgent and worthwhile conservation activity.

Obviously enormous damage has been done to many fish habitats and the situation is often not easy to reverse – especially in the short term where fish species or communities are severely threatened. In many cases, unique stocks have completely disappeared. Even where habitat restoration is contemplated, stock transfer (discussed below) could be an important interim measure. However, there are a number of important examples of habitat restoration in temperate areas and it should be emphasised that habitat protection and restoration are the principle long-term means through which successful fish conservation will be achieved.

Translocation of endangered stocks of fish can be done without any threat to the existing stocks, but it is important that certain criteria are taken into account in relation to any translocation proposal. With most fish it is possible to obtain substantial numbers of fertilised eggs by catching and stripping adult fish during their spawning period. The adults can then be returned safely to the water to spawn in future years. Fortunately, most fish produce an enormous excess of eggs and so substantial numbers can be taken at this time without harm. Having identified an appropriate water in which to create a new population, this can be initiated by placing the eggs there, or hatching the eggs in a hatchery and introducing the young at various stages of development.

In view of the urgency relating to a number of endangered populations of fish one of the most urgent tasks needing to be carried out is the development of techniques for handling these rare fish and the establishment of new safeguard populations in suitable sites. One of the most difficult aspects of programmes to date has been to locate sites, which are suitable ecologically, geographically and where the owner is sympathetic to the proposals. A common problem is that the otherwise suitable site is already being used as a fishery for other, common, species.

Captive breeding is widely used throughout the world for a variety of endangered animals, including fish. However, for most animals it can really only be regarded as a short term emergency measure, for a variety of genetic and other difficulties are likely to arise if small numbers of animals are kept in captivity over several generations or more.

However, short-term captive breeding involving only one generation does have some advantages for a number of species. It is especially relevant where translocations are desirable but it is difficult to obtain reasonable numbers of eggs or young because of ecological or logistic constraints. In such cases there are considerable advantages to be gained in rearing small numbers of stock in captivity and then stripping them to obtain much larger numbers of young for release in the wild. Because of genetic problems related to the 'bottleneck' effect and inbreeding it should not be carried out for more than one generation from the wild stock.

Modern techniques for rapid freezing of gametes to very low temperatures (cryopreservation) have proved successful for a variety of animals, including fish. After freezing for many years and then thawing the material is still viable. However, the technique is successful only for sperm and though much research is at present being carried out on eggs, no successful method of cryopreservation has yet been developed. The technique is therefore of only limited value in relation to the conservation of fish species.

However, where a particular stock seemed in imminent danger of dying out it would seem worthwhile giving consideration to saving at least some of its genetic material through the cryopreservation of sperm. When it is possible to preserve female gametes in a similar way, the technique will have obvious possibilities in relation to the short-term conservation of a wide variety of fish species.

An overall assessment of the present status of each species in Europe as a whole is given in the species accounts, using the IUCN (1994) categories of threat, which may be summarised as follows: **Extinct**: 'when there is no reasonable doubt that the last individual has died'. **Extinct in the Wild**: 'when it is known only to survive in cultivation, in captivity or as a naturalised population (or populations) well outside the past range.' **Critically Endangered**: 'when it is facing an extremely high risk of extinction in the wild in the immediate future'. **Endangered**: 'when it is not Critically Endangered but is facing a very high risk of extinction in the wild in the near future.' **Vulnerable**: 'when it is not Critically Endangered or Endangered but is facing a high risk of extinction in the wild in the medium-term future'. **Lower Risk**: 'when it has been evaluated, does not satisfy the criteria for any of the categories Critically Endangered, Endangered or Vulnerable'. **Data Deficient**: 'when there is inadequate information to make a direct, or indirect, assessment of its risk of extinction based on its distribution and/or population status.' **Not Evaluated**: 'when it has not yet been assessed against the criteria'.

Future trends

The pressures described above are likely to continue and to be supplemented by other complications as the demand for water by industrial, domestic, and recreational concerns increases. The main impact on fish would seem to be in favour of angled species in most cases, and extremely detrimental to non-sport fish – especially rare species. Generalisations are difficult to make, but the main trends appear to be as follows. 1 Natural, stable mixed fish communities are likely to be changed through poisoning (or shocking) followed by stocking with unstable, virtual monoculture populations of introduced sport species such as Rainbow Trout. 2 Eutrophication processes are likely to continue and to speed up the succession in natural waters towards conditions which favour coarse species. The fish which will be most successful here are likely to be coarse species of sporting value, such as Bream. 3 Pollution has eradicated certain populations of estuarine and migratory species but seems likely to lessen in future years, so that there is the possibility of re-establishing many of these fish. In some cases, e.g. in the Rivers Rhine and Thames, parts of the fish fauna are recolonizing naturally. Where pollution is a continuing or increasing problem, sensitive game species, such as Atlantic Salmon, will disappear before coarse species, such as Roach. 4 Obliteration of small water bodies will continue, with consequent destruction of populations of small species such as Mudminnows. 5 Haphazard introduction of various kinds will continue, mainly involving sport species or small species used as bait; but regardless of the ecological effects of such introductions the value of the gene-pools within the species concerned must deteriorate as a result of indiscriminate mixing and dominance by the common types. 6 The general impact of increased land use, further connections between catchments, etc., is likely to favour the distribution and abundance of the more adaptable species, to the disadvantage of the more sensitive forms with poor powers of dispersal. Examples of the former are Ruffe and Dace, and of the latter, Vendace and Arctic Charr.

Although there may be an increase in the abundance of sport species throughout Europe, from most other points of view there will be a loss within the resource as a whole. Among sport species this is likely to involve loss of genetic material, increased incidence of disease and parasites, and loss of stability within the ecosystems involved. Among other, non-sport, species these same changes may be expected, along with the loss of many small local populations; in extreme cases, unique taxa may disappear.

The main hope for the future would seem to lie in persuading those concerned in sport fishery management to adopt a more responsible attitude towards our native fish stocks, and in stimulating those concerned with all aspects of water use to take account of the value of this resource in their planning. In particular, the following lines of action would seem to be worth developing.

Firstly, increasing the awareness, among those concerned, of the conservation value of fish species and populations in their geographic area or region of control. Secondly, slowing down the obliteration of small aquatic systems and indeed increasing the number available by the creation of new ponds within country parks, nature reserves, etc. Such new ponds, and indeed all new reservoirs, should be designed so that the ecosystem developing within them is suitable for various fish species. Thirdly, there should be more stringent control of poisoning of waters and of the movement of fish stocks (even of common species) from one catchment to another. Lastly, there should be further active work on the conservation of rare species, and of valuable stocks of common species, by protecting the ecosystems in which they occur at present and by transferring selected stock to appropriate waters in the same area. Even with common species, there should be action to reintroduce species to waters (including estuaries) from which they have disappeared, if conditions there currently appear suitable, and to establish appropriate mixed communities in new water bodies. The stocks used for such introductions should be chosen carefully.

Also, it should not be thought necessary that every body of water has a fish population. It is highly desirable that, among the diversity of water bodies in Europe, there should remain a number which have no fish species, thus allowing for the development there of the interesting natural communities of aquatic plants and invertebrates which exist in the absence of fish.

Classification

The classification and naming of living creatures has always appeared something of a mystery to the layman. Much of the misunderstanding and difficulty has arisen over the fact that biologists have used Latin or Greek derivations for the scientific names chosen. This, together with ignorance over the status of specific, generic, and other higher rank names, has led to a division between scientists who only recognise the two- or three-worded scientific name for a species wherever it occurs, and laymen who may use different names for the same species in different countries, and even in different parts of the same country.

A good example of the complexity of naming fish is found in the Northern Pike, which is widespread in Europe, Asia, and North America. Biologists all over the world clearly recognise this fish by the scientific name *Esox lucius*. In Europe alone, however, the following common names are used in different countries (and even within these countries various other local dialect names are found): Czechoslovakia Stika obecna; Denmark Gedde; Germany Hecht; Great Britain Pike; Spain Lucio; France Brochet; Ireland Lius; Yugoslavia Stuka; Italy Luccio; Hungary Csuka; Netherlands Snoek; Norway Gjedde; Poland Szczupak; Rumania Stiuca; Russia Shtschuk; Finland Hauki; Sweden Gadda; Turkey Turna baligi. In North America, *Esox lucius* is known variously as Redfin Pickerel, Banded Pickerel, Trout Pickerel, Grass Pickerel, Mud Pickerel, Bulldog Pickerel, Pickerel, Northern Pike and Red-finned Pike. The value of a single international name for each species is quite clear in this context!

The two-worded scientific name for a species is equivalent to the names given to many individual humans, with the first, generic, name (e.g. *Esox*) equivalent to the surname, and the second, specific, name (e.g. *lucius*) equivalent to the christian name. Although of course they are used in reverse here. A species name is often followed by the name of the scientific authority who first described it, and the year the description was published (viz. *Esox lucius*, Linnaeus, 1758). The scientific name chosen for each species usually refers to some character of shape or habit, but occasionally species are named after particular people or places. Where a subspecies is recognised there is a third, subspecific, name – e.g. *Esox lucius bergi*.

Taxonomy is the study of the various characteristics of plants and animals, how they differ from and relate to each other and how they should be properly named in the scientific literature. There are complex international rules for doing this. Unfortunately, for fish and many other groups, there is still much work to be done and the taxonomic status and scientific name of many species is still in a fluid state. Largely because of insufficient information, on a large enough geographic scale, taxonomists often do not agree with each other and confusion exists. A recent and thought-provoking publication by Kottelat (1997) has reviewed the situation regarding European freshwater fish and should be read by anyone with an interest in this topic. The first edition of the present Hamlyn book included 215 species. This edition includes 316 species. The difference is because in the 23 years intervening between the two editions, many new species have been described and the status of existing ones revised. However, Kottelat has suggested that there are actually 358 species in Europe! This difference in opinion regarding numbers exists largely because Kottelat maintains that many of the original species described are still valid and does not accept the recent revisions and 'lumping' of species, which is often based on modern methods such as DNA analysis, etc. Some of these doubtful 'species' are mentioned in this book but are not dealt with in full, being better regarded as subspecies or local races.

The naming of species is not, however, merely a haphazard process subject to the whim of each scientist discovering a new one. The whole basis of the way in which names are grouped together is founded on belief in the theory of evolution which suggests that living things are continually changing to meet the demands of their environment (see fig. 27). Thus a single species, subjected to differing stresses in two different waters can, over a long period of time, evolve into two new species which are distinctly different from the original stock, and eventually so different from each other that they are incapable of interbreeding. There is still a great deal of research to be done in this important field and Kottelat (1997) has done science a great service in pointing out the problems and deficiencies which exist in Europe.

Different though they are, however, two species which have many features in common may be more closely related to each other than to any other fish and it is likely that they would be placed in the same genus. Thus in North America there is a second species of pike known as the Muskellunge. Its scientific name is *Esox masquinongy*. (European pike anglers will be interested to know that this fish grows up to 1.8 m (6 ft) long and over 50 kg (110 lb) in weight). Together with several other species, these two species of *Esox* make up the family Esocidae. This family and another related one, the Umbridae, make up a small order known as Haplomi. There are many other, mostly larger, orders of fish which are united with the Haplomi in the Class Osteichthyes. All members of this class possess a bony skeleton, in contrast to members of the only other class of fish found in European fresh waters, the Marsipobranchii. Here the skeleton consists only of soft cartilage.

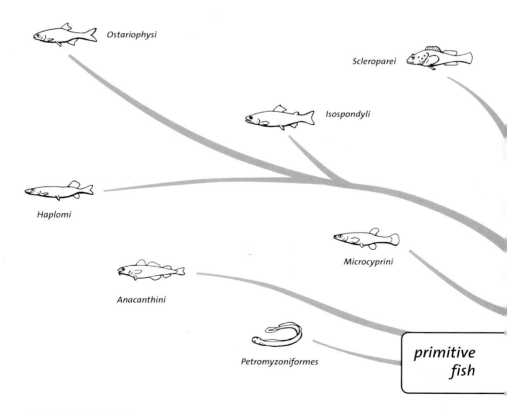

Fig. 27 Possible evolutionary
relationships among the orders of
European freshwater fish

It will be seen, therefore, that different fish species are classified together in increasingly larger groups, the relationships within each group being believed to be due to common ancestry. The classification covering European freshwater fish is given in the check list below, and as a final example the status of pike and man – both vertebrates – is given here in terms of their classification.

	PIKE	**MAN**
Sub-phylum	Vertebrata	
Class	Osteichthyes	Mammalia
Order	Haplomi	Anthropoidea
Family	Esocidae	Hominidae
Genus	*Esox*	*Homo*
Species	*lucius*	*sapiens*

Some of the important characters of the various European fish are indicated in the identification keys and descriptions of families, genera and species which are given later in this book.

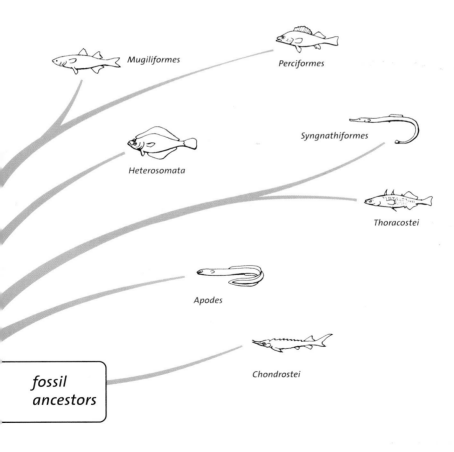

Mugiliformes

Perciformes

Syngnathiformes

Heterosomata

Thoracostei

Apodes

Chondrostei

*fossil
ancestors*

Key to families of European freshwater fish

1 No paired fins; 7 pairs of gill openings; no lower jaw, mouth
   in adults a sucking disc; a single median nostril between the eyes   **Petromyzontidae** (page 62)

   1 or 2 paired fins; 1 pair of gill openings, each protected
   by an operculum; lower jaw present, mouth never
   a sucking disc; paired nostrils anterior to eyes   *Go to 2*

2 Body covered with bony rings; snout tubular; caudal fin absent
   or very small   **Syngnathidae** (page 210)

   Body not covered with bony rings; snout not tubular;
   caudal fin always present and more or less normal   *Go to 3*

3 Upper lobe of caudal fin much longer than lower (heterocercal);
   5 longitudinal rows of large bony plates on body;
   snout greatly elongated   **Acipenseridae** (page 70)

   Caudal fin more or less symmetrical (holocercal); no large
   bony plates on body; snout normal   *Go to 4*

4 Small fleshy protuberance over eye; single dorsal fin extending
   from head almost to tail fin   **Blenniidae** (page 232)

   No fleshy protuberance over eye; dorsal fin or fins shorter   *Go to 5*

5 1 dorsal fin, or if 2, then posterior (adipose) fin small and fleshy,
   without rays; pelvic fins, where present, approximately midway
   between pectoral fins and anus, or near anus; pneumatic
   duct present between swim-bladder and oesophagus   *Go to 6*

   2 dorsal fins, or if 1, then either this is divided into 2 distinct parts,
   the anterior part being very spiny or replaced by isolated spines,
   or the body is greatly flattened with both eyes on one side of the
   head; pelvic fins just below, or slightly posterior to, pectoral fins;
   pneumatic duct absent between swim-bladder and oesophagus   *Go to 22*

6 Barbels present on head, largest pair longer than pectoral fins, dorsal fin with less than 8 rays; scales absent

Barbels either absent on head, or if present then much shorter than pectoral fins; dorsal fin with more than 8 rays; scales present (although very small in 2 families)

Go to **8**

---

7 4 or 6 barbels, the dorsal pair very long, reaching back to the dorsal fin, anal fin very long with about 90–92 rays

Siluridae (page 166)

8 barbels, the dorsal pair short, not reaching back beyond the operculum; anal fin short with only 20–22 rays

Ictaluridae (page 164)

---

8 2 dorsal fins, posterior (adipose) fin fleshy without rays

Go to **9**

1 dorsal fin

Go to **12**

---

9 Scales relatively small, more than 100 along lateral line; red pigment often present in skin

Salmonidae (page 177)

Scales relatively large, less than 100 along lateral line; red pigment rarely present in skin

Go to **10**

---

10 Lateral line complete almost to caudal fin; teeth absent or poorly developed; pelvic axillary process present

Go to **11**

Lateral line present only for about first 10–20 scales; teeth well developed; pelvic axillary process absent

Osmeridae (page 170)

---

11 Dorsal fin large, depressed length of which is much greater than that of head, and with more than 20 rays; large black pigment spots normally present in skin; small teeth present

Thymallidae (page 189)

Dorsal fin normal, depressed length of which is never greater than that of head, and with less than 20 rays; large black pigment spots never in skin although small black chromatophores may be common

Coregonidae (page 171)

12 Dorsal fin distinct from caudal fin; pelvic fins present; body not extremely elongate   *Go to* **13**

Dorsal fin continuous with caudal and anal fins; pelvic fins absent; body extremely elongate   **Anguillidae** (page 76)

---

13 Scales on ventral surface keeled; lateral line absent; large elongate scales over inner part of caudal fin   **Clupeidae** (page 78)

Scales on ventral surface not keeled; lateral line present; no elongate scales over inner part of caudal fin   *Go to* **14**

---

14 Dorsal fin mostly posterior to anus; teeth present   *Go to* **15**

Dorsal fin entirely or mostly anterior to anus; teeth absent   *Go to* **20**

---

15 Head elongate, mouth very large with well-developed canine teeth; lateral line present   **Esocidae** (page 168)

Head and mouth normal with small teeth; no lateral line   *Go to* **16**

---

16 Anal fin with less than 8 rays; 13–16 dorsal rays   **Umbridae** (page 168)

Anal fin with more than 8 rays; 9–13 dorsal rays   *Go to* **17**

---

17 Anal fin in male normal; oviparous; dorsal fin with more than 9 rays; adult teeth trifid at tip   *Go to* **18**

Anal fin in male produced to form copulatory organ; viviparous; dorsal fin with less than 9 rays; adult teeth conic at tip   **Poeciliidae** (page 204)

---

18 Lateral line with 33–38 scales; 18–19 pectoral rays   **Fundulidae** (page 199)

Lateral line with 25–32 scales; 14–15 pectoral rays   *Go to* **19**

19 Teeth unicuspid and conical; 12–14 rays in anal fin; 29–32 scales on lateral line

Valenciidae (page 200)

Teeth tricuspid; 9–12 rays in anal fin; 25–30 scales on lateral line

Cyprinodontidae (page 202)

---

20 Less than 5 barbels on head; mouth normal; scales usually distinct on body

Cyprinidae (page 86)

More than 5 barbels on head; mouth small; scales on body indistinct

Go to 21

---

21 Bifid spine in a pocket under each eye; 3 pairs of short barbels, all short OR no bifid spine; 5 pairs of long barbels

Cobitidae (page 154)

Bifid spine absent or very undeveloped; 3 pairs of barbels, one pair longer than the others

Balitoridae (page 162)

---

22 2 dorsal fins, or if 1, then divided into 2 distinct parts; the anterior part being very spiny; body never greatly flattened or with isolated spines

Go to 23

1 dorsal fin; body greatly flattened or with a row of dorsal spines

Go to 32

---

23 Head with single barbel below mouth and 1 small barbel beside each nostril; anal fin with more than 60 rays

Gadidae (page 190)

Head without barbels; anal fin with less than 60 rays

Go to 24

---

24 Well-developed scales over most of body; anterior dorsal fin rays rigid

Go to 25

Scales absent over most of body; anterior dorsal fin rays flexible

Cottidae (page 213)

25 Lateral line absent; less than 5 spiny rays in anterior dorsal fin; dorsal fins widely separated, distance between them always exceeding length of longest dorsal ray

Go to **26**

Lateral line present; more than 5 spiny rays in anterior dorsal fin; dorsal fins continuous or close together, distance between them never exceeding length of longest dorsal ray

Go to **27**

26 Head flattened dorso-ventrally, scaled dorsally; anal fin with less than 10 rays

Mugilidae (page 191)

Head compressed laterally, not scaled dorsally; anal fin with more than 10 rays

Atherinidae (page 196)

27 1 nostril on each side of head; less than 30 scales along lateral line

Cichlidae (page 231)

2 nostrils on each side of head; more than 30 scales along lateral line

Go to **28**

28 2 or fewer anal spines present; either less than, or more than, 9 or 10 spiny rays in first dorsal fin

Go to **29**

3 or more anal spines present; 9 or 10 spiny rays in first dorsal fin

Go to **31**

29 2 anal spines present; tail fin forked; more than 10 spiny rays in first dorsal fin; pelvic fins not joined medially

Percidae (page 222)

Anal spines absent; tail fin rounded; less than 9 spiny rays in first dorsal fin

Go to **30**

30 Pelvic fins joined medially

Gobiidae (page 233)

Pelvic fins not joined medially

Eleotridae (page 248)

31 Second dorsal fin jointed to first anteriorly; less than 70 scales
along lateral line; anal fin convex

Centrarchidae (page 217)

Second dorsal fin separated from first anteriorly; more than 70
scales along lateral line; anal fin concave

Moronidae (page 215)

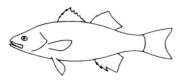

32 3 or more strong spines anterior to dorsal fin; body not flattened;
eyes on either side of head; pelvic fins with less than 3 rays

Gasterosteidae (page 206)

No spines anterior to dorsal fin; body extremely flattened
with both eyes on one side of the head (usually the right);
pelvic fins with more than 3 rays

Pleuronectidae (page 248)

Lamprey Family
Petromyzontidae

The Petromyzontidae, or lampreys, belong to a small but important group known as Agnatha – literally 'jawless' fish. Thus they are quite distinct from all the other fish described in this book which have upper jaws fixed closely to the skull and hinged lower jaws which oppose them. The lampreys, in contrast, have no lower jaws and, in adults, the whole mouth is surrounded by a round sucker-like disc within which are developed horny rasping teeth – well developed in parasitic species, poorly formed in non-parasitic species, which do not feed as adults. These teeth vary in position and number among the species, and are an important aid to identification.

Lampreys also have a number of other very characteristic features: they are always very eel like in shape, but have no paired fins or scales. They have no bones – all the skeletal structures are made up of stiff but flexible cartilage. There is a single nostril, situated on top of the head, just in front of the eyes; the latter may not be functional or even visible in the young. The gills open directly to the sides of the head (i.e. there is no operculum), so that there is a lateral row of gill pores immediately behind each eye. The adults have two dorsal fins which are often more or less continuous with the elongate tail fin.

Most lampreys have a similar life cycle, which involves the migration of adults from their feeding grounds to special spawning areas – normally stony or gravelly stretches of running water. There they spawn in pairs or groups, and the eggs are laid in crude nests, just a depression in the gravel, created by the adults lifting away small stones with their sucker mouths. These stones are used to surround, and sometimes to cover and protect, the eggs, while the nest itself may often be under a large stone, log, or clump of vegetation. After hatching, the elongate larvae, known as ammocoetes, swim or are washed by the current to areas of sandy silt in still water where they burrow and spend the next few years in tunnels. They are blind, the sucker is incomplete around the mouth and the teeth are undeveloped. The *ammocoete* larvae feed by creating a current which draws loose organic particles (coated with bacteria), and minute plants such as diatoms, into the pharynx. They there become entwined on a slimy mucous string which is swallowed regularly by the animal. The change from larval to adult form is a fairly dramatic metamorphosis which takes place within a relatively short time – usually just a few weeks. The mouth develops into a full sucker inside which are rasping teeth. The skin becomes much more silvery and opaque except over the eyes, where it clears to give the animal proper vision for the first time. The adults then migrate, usually downstream, away from the nursery areas.

Some species of lampreys never feed as adults – after metamorphosis these migrate upstream, spawn and die. However, after metamorphosis most species are parasitic, living on various fish which they attack in rivers, large freshwater lakes or in the sea, where most of the adult life is spent. They attach themselves to the sides of fish and rasp away the skin,

Oral disc of an adult lamprey, showing main mouthparts

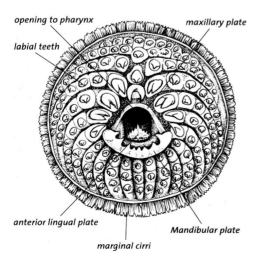

opening to pharynx

labial teeth

maxillary plate

anterior lingual plate

Mandibular plate

marginal cirri

62

Key to Petromyzontidae in Europe

1	Maxillary plate narrow, bearing 1 or 2 teeth	*Go to* **2**
	Maxillary plate wide, bearing 1 tooth on each side	*Go to* **3**
2	2 teeth on maxillary plate; 7–8 teeth on mandibular plate; teeth strong, sharp; anterior lingual plate with median depression	***Petromyzon marinus*** p69
	Normally 1 tooth on maxillary plate; 5 teeth on mandibular plate; teeth weak, blunt, rounded; anterior lingual plate without median depression	***Caspiomyzon wagneri*** p64
3	Outer and inner labial teeth present; lower labial teeth present; mediolateral labial teeth bifid	*Go to* **4**
	Outer lateral labial teeth absent; mediolateral labial teeth bifid or trifid	*Go to* **7**
4	Caudal fin rather squared; pigment absent on upper and lower lips; more than 60 trunk myomeres	*Go to* **5**
	Caudal fin rounded; upper and lower lip strongly pigmented; less than 60 trunk myomeres	***Eudontomyzon hellenicus*** p65
5	Body thickest near middle; lingual plate with 9–13 teeth; uniformly coloured – no spots or mottling	***Eudontomyzon danfordi*** p64
	Body thickest anteriorly; lingual plate with 9 or less teeth	*Go to* **6**
6	Lingual plate with 5–9 teeth	***Eudontomyzon vladykovi*** p66
	Lingual plate with 5 teeth; always well spotted or mottled	***Eudontomyzon mariae*** p65
7	Mediolateral labial teeth bifid	*Go to* **8**
	Mediolateral teeth trifid	*Go to* **9**
8	Oral teeth well developed, labial teeth strong and pointed; 65–77 trunk myomeres	***Lethenteron camtschaticum*** p67
	Oral teeth poorly developed, labials minute and blunt; 52–60 trunk myomeres	***Lethenteron zanandreai*** p68
9	Mandibular plate with 7–10 teeth, all of them strong and sharp; dorsal fins separate	***Lampetra fluviatilis*** p66
	Mandibular plate with 5–9 teeth, all of them weak and blunt; dorsal fins connected	***Lampetra planeri*** p67

eating both it and the body fluids and muscle beneath. On reaching sexual maturity, the adult lampreys migrate back again to the spawning areas. At spawning time they are very vulnerable to predators and they form an important part of the food of piscivores such as herons, gulls, otters and mink.

The prey often never recovers from a lamprey attack (especially if the body cavity is penetrated) and in some waters lampreys are a serious pest to commercial fish stocks. The most famous example of this is in North America, where canalisation gave the Sea Lamprey access, for the first time, to the Great Lakes. Various commercial fish stocks there were soon seriously depleted, particularly the American Lake Charr whose population collapsed in a dramatic way. For many years the Great Lakes Fishery Commission has spent about $10 million per annum controlling the numbers of Sea Lamprey there. In contrast, in Europe, lampreys are still caught commercially in a number of countries (e.g. Sea Lampreys in Portugal and River Lampreys in Finland), and in some are regarded as a great delicacy. River Lampreys, smoked or soaked in oil, were formerly highly prized in England, and King Henry I is supposed to have died from a surfeit of them – probably due to the toxins which occur in the skins of most species. Those who know how to cook lampreys drop them live into salt beforehand – the extensive amounts of mucus which are exuded at this time eliminate most of the toxins.

Lampreys occur mainly in the cool temperate zones of the North and South Hemispheres. There are 40 species of Petromyzontidae world-wide: 10 of these occur in fresh water in Europe, where there are also a number of subspecies.

Caspian Lamprey *Caspiomyzon wagneri*

Size 20–40 cm; maximum 55 cm – this fish weighed 205 g. **Distinctive features** characteristic tooth pattern; adults of uniform (usually grey) colour. **Distribution** a native anadromous species widespread in the Caspian Sea and its larger, easily accessible rivers, especially the Volga. **Reproduction** March–May. Eggs laid in nests among stones and gravel in flowing water. The ammocoete larvae, which live buried in sand and silt, metamorphose after about 3 years and descend to the sea. 20,000–32,000 eggs per female. **Food** filtered organic material (algae and detritus) when larvae. The adults probably feed on other fish, but some have been found to contain vegetable material. **Value** an important commercial species caught in nets and traps, mainly during spawning migrations (e.g. in the Volga). Early this century, 15–30 million fish were caught annually in the lower Volga and used as food. Of no sporting significance. **Conservation Status** Vulnerable.

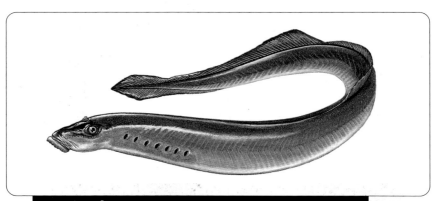

Danube Lamprey *Eudontomyzon danfordi*

Size 15–25 cm; maximum 30 cm. **Distinctive features** thickest part of body near middle; 9–13 teeth on lingual plate – central cusps largest. **Distribution** found only in the Danube basin or waters closely associated with it. A purely freshwater species. **Reproduction** April–May, mostly in the smaller tributaries of the Danube. Larvae live in silted sandy beds for 3–4 years before metamorphosing into adults which move down into the larger rivers. The adults mature after about 1 year, when they spawn and die. **Food** small organic particles filtered by larvae. The adults feed on various other stream fish. **Value** of no commercial or sporting value. **Conservation Status** Vulnerable.

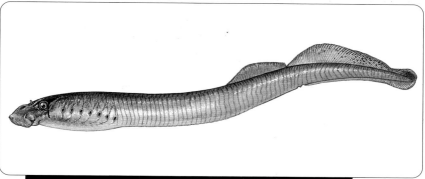

Greek Lamprey *Eudontomyzon hellenicus*

Size 12–14 cm as larvae, maximum 19 cm; a larva 189 mm long weighs 12.3 g, an equivalent adult would be 159 mm and 7.4 g **Distinctive features** characteristic mouth structure and tooth pattern; rounded caudal fin. **Distribution** a purely freshwater species, found only in the Strymon (Aegean drainage) and Louros (Ionian drainage) River basins in Greece where it is regarded as critically endangered. **Reproduction** in January and in May, in clean karstic springs and the head waters of rivers. Larvae live in silt beds downstream for about 4–6 years before metamorphosing into adults which live for only 3–4 months. **Food** larvae feed on fine organic detritus and algae. The adults do not feed. **Value** of no commercial or sporting value but protected in Greece as an endangered species. **Conservation Status** Critically Endangered.

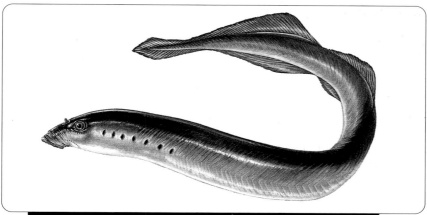

Ukrainian Lamprey *Eudontomyzon mariae*

Size 14–20 cm; maximum 21 cm. **Distinctive features** numerous small outer labial teeth along the margin of the disc, usually in several rows; lower labial teeth present but rarely forming a continuous patch. *Eudontomyzon stankokaramani* Karaman 1974 may be a subspecies. **Distribution** occurring only in fresh water, in rivers entering the northern shore of the Black Sea (e.g. Don, Dnieper and Dniester). **Reproduction** March–May, mainly in smaller streams among stones and gravel. Larvae live in silt beds of streams and rivers for 4–6 years before metamorphosing into adults which spawn soon afterwards and then die. 2,429 eggs were found in a female 14.7 cm long. **Food** larvae filter fine organic particles from their surroundings. The adults do not feed. **Value** of no commercial or sporting value. **Conservation Status** Vulnerable.

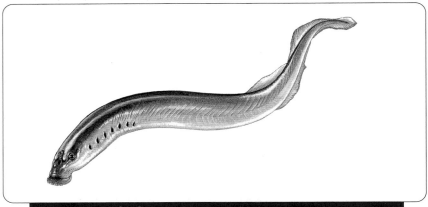

Vladykov's Lamprey *Eudontomyzon vladykovi*

Size 15–20 cm; maximum 21 cm. **Distinctive features** body thickest anteriorly; 5–9 teeth on lingual plate; innermost lateral teeth usually with 2–4 points. **Distribution** a purely freshwater species found only in parts of the Danube basin. **Reproduction** April–May, mainly in small tributaries. Larvae live in silted areas of streams for 4–6 years before metamorphosing into adults which spawn soon afterwards. **Food** fine filtered organic particles when larvae. The adults do not feed. **Value** of no commercial or sporting value. **Conservation Status** Vulnerable.

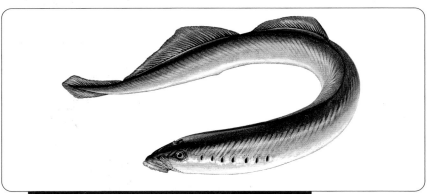

River Lamprey *Lampetra fluviatilis*

Size 30–35 cm; maximum 50 cm and 150 g; the mean weight of females 32–34 cm long is about 62 g. **Distinctive features** maxillary plate wide, but no lower labial teeth; 7–10 teeth on mandibular plate. **Distribution** occurs over much of western Europe from southern Norway to the western Mediterranean, in the sea and in accessible rivers. There are a number of purely freshwater populations (e.g. in Loch Lomond and Lake Ladoga). **Reproduction** April–May. Eggs laid in nests among stones in running water. The ammocoete larvae live in silted stream beds for 3–5 years, then metamorphose at about 10–12 cm into young adults which migrate to the sea (or a large lake). After 12–18 months, the mature adults migrate upstream again in winter and spring. 19,000–20,000 eggs per female. All die after spawning. **Food** filtered organic material when larvae, fish when adult. **Value** an important commercial species in Scandinavia, the adults being caught in nets and traps as they migrate upstream. Many are smoked or grilled and eaten whole – for the gut is empty during migration and there is no bone. **Conservation Status** Vulnerable.

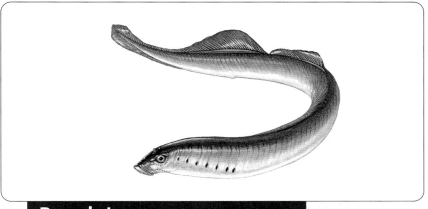

Brook Lamprey *Lampetra planeri*

Size 12–15 cm; maximum 21 cm. Fully grown ammocoetes are larger than the adults. **Distinctive features** wide maxillary plate present but no lower labial teeth; 5–9 blunt teeth in mandibular plate. **Distribution** a purely freshwater species occurring in streams in much of western Europe, especially in basins associated with the Baltic and North Seas. **Reproduction** April–June. Eggs laid in nests in stones in running water. Larvae live in silt in streams and rivers, metamorphosing after 3–5 years to adults which migrate upstream, spawn and die soon afterwards. 854–1,400 eggs per female. **Food** filtered organic material when larvae. The adults do not feed. **Value** of no commercial value, but the larvae are occasionally used as bait for angling. **Conservation Status** Vulnerable.

Arctic Lamprey *Lethenteron camtschaticum*

Size 40–60 cm; maximum 63 cm; lampreys with a mean length of 54 cm weigh about 146 g. **Distinctive features** a wide maxillary plate with well-developed lower labial teeth. **Distribution** northern Europe, Asia and North America in rivers associated with the Arctic and Pacific basins. **Reproduction** May–June, in rivers easily accessible from the sea. Larvae live in silted river beds for 3–4 years before metamorphosing and migrating to the sea where they grow to maturity. 80,000–107,000 eggs per female. **Food** fine organic particles when larvae, fish (including anadromous species) when adult. **Value** little used commercially and of no sporting value. **Conservation Status** Vulnerable.

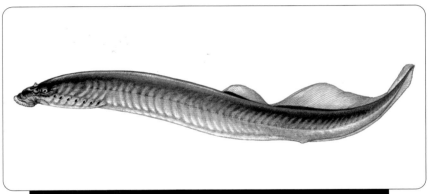

Lombardy Lamprey *Lethenteron zanandreai*

Size 10–20 cm; maximum 22 cm; a larva of 21.6 cm gives rise to an adult of 20.0 cm. **Distinctive features** 2 cusps on median tooth of lateral circumoral row; the mean number of trunk myomeres is 55. **Distribution** found only in Italy in the Po drainage basin, especially the Saluzza, Villafranca and Bielese areas. Occurs in cold water springs in the foothill zone, where the bottom is sand or mud. **Reproduction** January–March among gravel beds. Larvae are 35–50 mm after one year; the larval life is about 4–5 years; that of adults about 6–8 months. Females lay about 2,000 eggs. Neotenous females have been recorded. All die after spawning. **Food** a non-parasitic species so the adults do not feed; the larvae are filter feeders on algae and detritus. **Value** of no commercial value but the larvae are sometimes used as bait. **Conservation Status** Critically Endangered.

adult

ammocoete larva

Sea Lamprey *Petromyzon marinus*

Size 50–70 cm; maximum 86 cm; a 53 cm specimen weighs about 360 g. **Distinctive features** characteristic tooth pattern; adults with very mottled colour patterning; the largest European lamprey. **Distribution** a native anadromous species common in most of the Atlantic coastal area of western Europe and eastern North America and in estuaries and easily accessible rivers there. **Reproduction** March–May. Clear eggs, laid in nests among stones in running water, hatch into elongated blind ammocoete larvae which live buried in silt in quiet waters for 2–5 years. They then metamorphose and migrate to the sea where they mature, returning to fresh water to spawn after 3–4 years. 34,000–240,000 eggs per female. **Food** filtered organic material (diatoms, bacteria, etc.) when larvae, fish when adult. **Value** formerly more popular than now, it is still trapped or netted commercially in some European rivers when ascending to spawn (e.g. in Portugal). Never angled for. A serious pest to commercial and sport fisheries in the Great Lakes of North America. **Conservation Status** Vulnerable.

Sturgeon Family
Acipenseridae

The Acipenseridae, or sturgeons, are an extremely interesting family of large primitive fish which are quite distinct from all other living bony fish. They occur only in the Northern Hemisphere, and the family contains about 23 species. The elongate body has no scales, but has several rows of characteristic bony plates which are often an important aid to identification. These plates become smoother with age and in some cases disappear altogether in old specimens. The tail is upturned into the dorsal lobe of the tail fin (a condition known as heterocercal) and forms its main support. The snout is elongate and projects well in front of the ventral mouth, anterior of which are four sensitive barbels; the exact form of these varies among the different species. The mouth itself is unusual in being a protrusible tube, well adapted for the mode of feeding, which is mainly on benthic invertebrates.

Sturgeons usually spawn in twos or threes in suitable stretches of large rivers or sometimes in lakes. The adults, which are often extremely large, are not particularly efficient swimmers; they are rarely found in fast-flowing, or small bodies of water. The eggs, which are rather unusual in being very dark in colour, are scattered loosely at spawning on to the bottom where, being adhesive, they stick to stones and logs. After hatching, the young migrate gradually to their nursery areas – usually large lakes, the lower reaches of large rivers or, in many species, the sea. There, growth is very slow and it may be up to 15 years before maturity is reached and the fish migrate back to their spawning areas. Some species can live for over 50 years.

Sturgeons have been highly valued as commercial species for hundreds, if not thousands, of years. Their large size, ease of capture, and tasty flesh and eggs have led to the development of important fisheries in almost all parts of the world where they occur. However, their slow rate of growth has rendered them very susceptible to overfishing, and a number of populations have become extinct because of this. Others have disappeared due to major pollution or barriers (e.g. weirs) to migration in the lower reaches of rivers which they once frequented. The flesh is eaten fresh or smoked, and in some areas is dried. The roe is eaten as caviar. To prepare this the ovaries are removed from ripe females, the eggs are separated from ovarian tissue, cleaned carefully and then packed in brine.

Ventral view of the head and mouth of the Ship Sturgeon

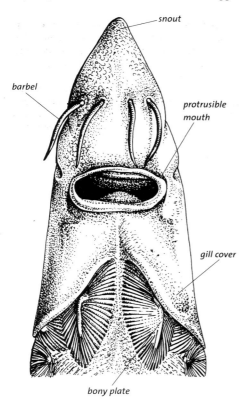

snout

barbel

protrusible mouth

gill cover

bony plate

Key to Acipenseridae in Europe

1	Mouth large, crescentic; branchiostegal membranes coalesced, forming a free fold below isthmus	***Huso huso*** p75
	Mouth small, transverse; branchiostegal membranes attached to isthmus, not forming a fold underneath	*Go to* **2**
2	Lower lip continuous, not interrupted medially; barbels fimbriate	***Acipenser nudiventris*** p73
	Lower lip with a median space	*Go to* **3**
3	Snout short, usually less than 60 per cent of head length	*Go to* **4**
	Snout elongate, usually more than 60 per cent of head length; lateral scutes 26–43	***Acipenser stellatus*** p74
4	Gill rakers fan-shaped, terminating in several tubercles. Lateral scutes 42–47	***Acipenser baeri*** p72
	Gill rakers not fan-shaped, terminating in a single tip	*Go to* **5**
5	More then 50 lateral scutes; barbels with distinct fimbria	***Acipenser ruthenus*** p74
	Less than 50 lateral scutes; barbels with no, or small, fimbria	*Go to* **6**
6	Snout shorter and rounded; lateral scutes 21–50	***Acipenser gueldenstaedtii*** p72
	Snout longer and pointed	*Go to* **7**
7	Lateral scutes less than twice as high as broad; snout less rounded when viewed from above	***Acipenser sturio*** p75
	Lateral scutes more than twice as high as broad; snout more rounded when viewed from above	***Acipenser naccarii*** p73

Depleted stocks in some areas have led to successful fish farming with a number of species. The fact that several of these interbreed to produce hybrids has helped make the industry successful. Several of these hybrids occur in nature, the most common of these being: *Huso huso* x *Acipenser stellatus*; *Huso huso* x *Acipenser nudiventris*; *Huso huso* x *Acipenser gueldenstaedtii*; *Acipenser nudiventris* x *Acipenser stellatus*; *Acipenser ruthenus* x *Acipenser sturio*; *Acipenser ruthenus* x *Acipenser stellatus*; *Acipenser gueldenstaedtii* x *Acipenser stellatus*.

Some anadromous sturgeons have two seasonal forms: a spring form which ascends rivers in spring and spawns in spring or early summer of the same year; and a winter form which ascends rivers in late summer or autumn and spawns the following year, having spent the winter in the river. Eight species occur in Europe.

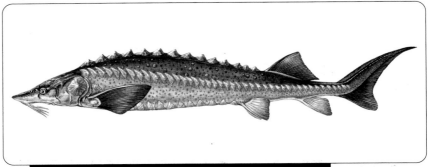

Siberian Sturgeon *Acipenser baeri*

Size 1.3–1.4 m; maximum 2 m (210 kg); specimens 110 cm weigh 8 kg. **Distinctive features** gill rakers fan-shaped, each ending in several tubercles; 42–47 lateral scutes. **Distribution** native to the large rivers of Siberia from the Ob to the Pechora; also in Lake Baikal; recent introductions to Lake Ladoga and the Baltic Sea, together with escapes from fish farms mean that it is caught regularly in western Europe. **Reproduction** May–June, in the main channels of large rivers over stony bottoms. Some fish are resident in rivers but others live in estuaries and migrate to spawn. The population in Lake Baikal migrates over 1,000 km up the River Selenga to its spawning grounds. Eggs hatch in 10 days at 17°C. Young descend to the sea after a few years and become mature at about 12 years (males) to 16 years (females). They can live for at least 60 years. The eggs of one female aged 26+ years, weighing 24 kg, weighed 5.8 kg and numbered 420,000. **Food** mainly bottom-dwelling invertebrates, especially midge larvae; in estuaries they eat crustaceans and worms. **Value** very important commercially, for farming and fisheries (e.g. River Yenisei), for both its flesh and its caviar. The most valuable fish in Siberia. **Conservation Status** Endangered.

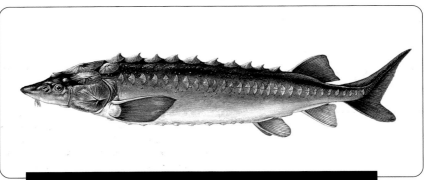

Russian Sturgeon *Acipenser gueldenstaedtii*

Size 1–2 m; maximum 2.3 m; maximum weight over 100 kg. **Distinctive features** short rounded snout; 24–49 lateral scutes. **Distribution** Black and Caspian Seas and their large rivers (e.g. Volga); some stocks are maintained by releases from fish farms. **Reproduction** May–June, in lower reaches of these rivers. After 3 years young descend to the sea, but a few stay in fresh water all their lives. Adults mature in 10–15 years and live up to 50 years. **Food** invertebrates (mainly insect larvae and molluscs) and some fish when in the sea. **Value** caught commercially in large numbers during migration. The roe is eaten as caviar. **Conservation Status** Endangered.

Adriatic Sturgeon *Acipenser naccarii*

Size 1–1.5 m; maximum 2 m. **Distinctive features** snout short and pointed; 32–44 lateral scutes more than twice as high as broad. **Distribution** Adriatic Sea and inflowing rivers (e.g. Po). **Reproduction** April–May in the middle reaches of rivers. The young then move down to the sea where they mature and migrate upstream again to spawn. **Food** bottom-dwelling invertebrates when young, invertebrates and fish when adult. **Value** formerly commercially important in the Adriatic area, but becoming rare and endangered. **Conservation Status** Endangered.

Ship Sturgeon *Acipenser nudiventris*

Size 1–1.25 m; maximum 2 m; specimens with a mean length of 1.5 m weigh 20 kg. **Distinctive features** a small mouth with a continuous lower lip; fringed barbels; 50–75 lateral scutes. **Distribution** common in the Black, Caspian and Aral Seas, and the larger rivers entering them (e.g. Kura and Syr-Darya). **Reproduction** April–May, in the middle and lower reaches of these rivers. The young move down to the sea where they take 10–15 years to mature. Some individuals may live up to 30 years, spawning several times during their lives. 216,000–1,002,000 eggs per female. **Food** invertebrates when young, invertebrates and some fish when adult. **Value** an important commercial species netted in very large numbers in the seas and rivers concerned. The eggs are eaten as caviar. **Conservation Status** Vulnerable.

Sterlet *Acipenser ruthenus*

Size the smallest sturgeon: 25–50 cm; maximum 92 cm – this fish weighed 19 kg. **Distinctive features** small mouth with median space in lower lip; barbels fringed; 55–70 lateral scutes. **Distribution** river basins entering the north of the Black Sea and east of the Caspian Sea and the western Baltic. This is mainly a freshwater species found in rivers and large lakes, but it is sometimes found in salt water. **Reproduction** April–June in fast flowing rivers over a gravel bottom. The adhesive eggs hatch in 6–9 days and the young gradually disperse downstream. Adults mature and start to spawn after 4–5 years, but may live up to 25 years. 7,800–76,400 eggs per female. **Food** mainly bottom-dwelling invertebrates – especially the larvae of insects such as blackflies, caddis flies and midges. **Value** important commercially and caught in nets or traps. Small specimens are commonly sold for aquaria. **Conservation Status** Vulnerable.

Stellate Sturgeon *Acipenser stellatus*

Size 1–1.5 m; maximum 190 cm. **Distinctive features** elongate snout; barbels short and smooth; 26–43 lateral scutes. **Distribution** northern Caspian Sea, Black Sea, Sea of Azov, and associated rivers; some stocks are maintained by the release of young from fish farms. Sometimes in the eastern Mediterranean. **Reproduction** April–June, in the middle reaches of rivers (e.g. Danube). The young move gradually down to the sea and mature in 8–12 years, living for up to 35 years. **Food** when young, invertebrates; when adult, invertebrates and small fish. **Value** caught commercially during migration. The flesh and roe (caviar) are both of importance. **Conservation Status** Endangered.

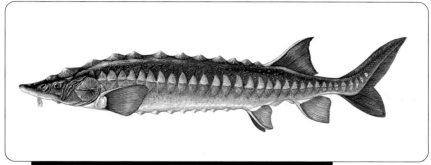

Atlantic Sturgeon *Acipenser sturio*

Size 1.5–2.5 m; maximum 3.45 m – this fish weighed 320 kg. **Distinctive features** snout pointed; lateral scutes (24–39) less than twice as high as broad. **Distribution** the coast of western Europe, including the Black Sea, and large accessible rivers. **Reproduction** April–July in middle reaches of rivers. The eggs adhere to stones and hatch after 3–5 days. The young move to the sea after 2 years, except for purely freshwater populations (e.g. in Lake Ladoga). Adults mature after 7–15 years and may live up to 50 years. 800,000–2,400,000 eggs per female. **Food** benthic invertebrates and, when adult, some fish. **Value** formerly important commercially and caught in large numbers for flesh and caviar. This species is now almost extinct and is one of Europe's most threatened fish. **Conservation Status** Endangered.

Beluga Sturgeon *Huso huso*

Size the largest sturgeon: 1.5–3 m; maximum 4.24 m; maximum weight 1220 kg. **Distinctive features** adults very large with large crescentic mouth; 37–53 lateral plates. **Distribution** Black and Caspian Seas and easily accessible rivers (e.g. Volga, Danube and Don); some stocks are now maintained by the release of young from fish farms. **Reproduction** April–May. The spawning grounds occur mainly in the lower and middle reaches of the rivers concerned. The young migrate downstream to the sea during their first year. They may take 15–20 years before maturing and entering the rivers again to spawn. Specimens as old as 60 years have been recorded. **Food** invertebrates when young, but mostly small species of fish (e.g. gobies and anchovies) when in the sea. **Value** of major commercial importance in the seas and rivers concerned. Adults of over 1,000 kg are not uncommon, and very heavy total catches are taken each year by the fisheries concerned. The females are especially valuable since they provide not only flesh, but their eggs are used as caviar. Beluga caviar is the most highly regarded of all caviars. **Conservation Status** Endangered.

Eel Family
Anguillidae

The Anguillidae, or eels, occur in Europe, Africa, Asia, North America and South America. The single genus contains 16 species. All are catadromous, moving early to fresh water and remaining there until maturity, then returning to the sea to breed. Relatively little is known about their spawning habits in the sea. Other eel families are mostly marine and there are many large species (e.g. Conger and Moray Eels), some of them found only in deep water.

The body shape is characteristic, being elongate and round in cross-section. Pelvic fins are absent. The eel larva (leptocephalus) is flattened from side to side and completely transparent and drifts in the plankton for long periods, growing until it metamorphoses into a miniature adult – known as an elver.

Only 1 species occurs in Europe, but at least one specimen of the American Eel *Anguilla rostrata* has been reported from Denmark.

yellow eel

silver eel

European Eel *Anguilla anguilla*

Size 40–90 cm; maximum about 2 m; British rod record 5.06 kg (1978). **Distinctive features** elongate, cylindrical body with small gill openings; 1 pair of pectoral fins, but no pelvic fins; minute scales embedded in the skin. **Distribution** the European coast (including the Black Sea) and in a variety of freshwater habitats accessible from the sea. **Reproduction** adults spawn in the Sargasso Sea. Larvae (leptocephali) drift across the Atlantic in the Gulf Stream for about 1 year before reaching Europe; they then migrate into fresh water as elvers, gradually maturing as 'yellow eels' at 8–18 years. These then migrate, as 'silver eels', back to the Sargasso Sea. Each female has several million eggs. **Food** invertebrates (especially molluscs and crustaceans) and fish. **Value** of commercial importance in many countries, where it is caught in traps, nets and by baited hooks. Often smoked. Of sporting value in a number of areas. Numbers have declined alarmingly in some areas in recent years. **Conservation Status** Lower Risk.

Herring Family
Clupeidae

The Clupeidae, many of which are known as herrings or shads, is a large family of mostly pelagic fish which are found in inshore sea areas and oceans all over the world. Most species are marine, but some are anadromous and a few live permanently in fresh water. There are several genera including a total of about 200 species.

The herrings are mainly small to medium-sized fish with a streamlined laterally compressed body covered by large, circular, cycloid scales. These are unusual in that the circuli are arched across the scale rather than arranged in concentric rings as in most cycloid scales. The ventral edge of the belly in most species is armed with scute-like scales, forming a toothed edge when viewed in profile. The head has large eyes with characteristic, fleshy eyelids. The mouth is usually terminal, and teeth are either small or absent. The many gill rakers lining the back of the pharynx are long and thin, and their number is an important character in identification. There is no lateral line.

Most members of the family are pelagic in habit and swim around in large shoals. These may comprise many thousands of individuals. The main foods of these shoals are the abundant masses of zooplankton which thrive in the richer parts of the sea. Feeding on these invertebrates is facilitated by the comb-like gill rakers, which help to separate the food from the water – the former being swallowed, the latter passing out through the gills. The herrings themselves form a major source of food for many larger fish, and for enormous colonies of sea birds in some parts of the world.

Because of the large size of the shoals, and the ease with which they may be captured, this is one of the most important families of commercial fish species in the world. They are caught mainly in gill nets, ring nets and trawls, and the total world catch is about 30 per cent of all fish caught by man. The flesh is particularly rich in fats and oils. Fish which cannot be marketed fresh are frozen, pickled or smoked. Fish which are too small for individual consumption are processed in bulk to make fish meal (for feeding to domestic mammals and birds) or for oil extraction.

Gills of two common Clupeidae

Since they are very dependent on the local abundance of zooplankton, the success of various populations of Clupeidae is closely linked with the fate of zooplankton, and thus in turn the phytoplankton, on which these small

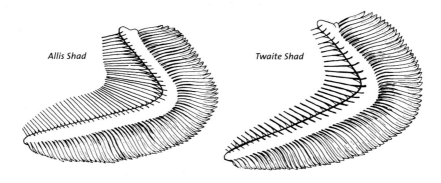

Allis Shad

Twaite Shad

Key to Clupeidae in fresh water in Europe

1	Upper jaw with a marked median notch; lower jaw articulating behind posterior margin of eye	*Go to* **2**
	Upper jaw with no, or only a small, median notch; lower jaw articulating anterior to posterior margin of eye	*Go to* **10**
2	Vomerine teeth absent	*Go to* **3**
	Vomerine teeth present	*Go to* **4**
3	More than 90 gill rakers on 1st arch; more than 70 lateral scales	***Alosa alosa*** p80
	Less than 60 gill rakers on 1st arch; less than 70 lateral scales	***Alosa fallax*** p81
4	More than 90 gill rakers on 1st arch	*Go to* **5**
	Less than 90 gill rakers on 1st arch	*Go to* **6**
5	Well developed teeth on vomer and palatine	***Alosa macedonica*** p81
	Poorly developed teeth on vomer and palatine	***Caspialosa caspia*** p82
6	Less than 45 gill rakers	*Go to* **7**
	More than 50 gill rakers	*Go to* **9**
7	Upper and lower profiles of head straight; gill rakers less than gill filaments in length	*Go to* **8**
	Upper and lower profiles of head rounded; gill rakers longer than gill filaments	***Caspialosa curensis*** p82
8	Eye large, its diameter more than 6 per cent of fish length; body depth more than 24 per cent of length	***Caspialosa saposhnikovi*** p84
	Eye small, its diameter less than 6 per cent of fish length; body depth less than 24 per cent of length	***Caspialosa maeotica*** p83
9	Usually 50–55 gill rakers; gill arches slender and weak; gill rakers half as long again as filaments	***Caspialosa suworowi*** p84
	Usually 60–90 gill rakers; gill arches thick and strong; gill rakers not exceeding filaments in length	***Caspialosa pontica*** p83
10	Body deep, more than 20 per cent of length; 52–64 gill rakers	***Clupeonella cultriventris*** p85
	Body shallow, less than 20 per cent of length; 38–52 gill rakers	***Clupeonella abrau*** p85

animals feed. The phytoplankton itself is dependent on an effective supply of nutrients brought up to the surface of the sea by up-welling currents. Thus the researches of hydrologists, who study and plot the ocean currents, are of great assistance in helping to understand the fluctuations in abundance of populations of Clupeidae.

European Clupeidae are mainly marine, but a few species are found in fresh water throughout their lives, and a number of others enter fresh water, either casually, or regularly to spawn at some time during their lives. Many of these anadromous species have declined in recent years due to pollution and physical barriers (e.g. weirs) in the rivers in which they spawn.

Altogether, 11 species occur in fresh water in Europe.

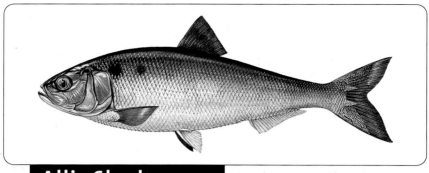

Allis Shad *Alosa alosa*

Size 30–50 cm; maximum 70 cm; British rod record 2.167 kg (1977). **Distinctive features** well-developed median notch in upper jaw; 85–130 long gill rakers; no vomerine teeth; more than 70 lateral scales. **Distribution** the coasts of western Europe from southern Norway to Spain and in the Mediterranean eastward to northern Italy. Migrates well upstream in large accessible rivers along these coasts (e.g. Loire, Garonne). **Reproduction** May–June, the clear eggs are laid over spawning gravels in flowing water during fast chases by males of females. The eggs hatch after 6–8 days and the young move downstream to the sea after 10–20 months. The adults mature after 3–4 years when they migrate, often in large numbers, back to the parent rivers. **Food** invertebrates, especially pelagic crustaceans, but also small fish. **Value** formerly more important commercially than now, it is still netted in some estuaries and the lower reaches of large rivers. **Conservation Status** Endangered.

Twaite Shad *Alosa fallax*

Size 25–40 cm; maximum 55 cm; British rod record 1.247 kg (1978). **Distinctive features** marked median notch in upper jaw; 30–80 short gill rakers; no vomerine teeth; less than 70 lateral scales. **Distribution** along most of the western European coast from southern Norway to the eastern Mediterranean, and in large accessible rivers along these coasts (e.g. Severn and Loire). Several purely freshwater populations, some of which have been given subspecific status (e.g. *Alosa fallax killarnensis* in Loch Leane, Ireland). **Reproduction** May–June, over gravel beds in the lower reaches and estuaries of these rivers. The eggs hatch after 5–8 days and the young migrate downstream to the sea over the next few months, except in purely freshwater populations where the young remain in large lakes (e.g. Leane, Como, Lugano and Maggiore). The adults mature after 3–4 years. 75,000–200,000 eggs per female. **Food** mainly invertebrates especially pelagic crustaceans. **Value** though less common than formerly, it is still netted commercially in some areas. Considerable numbers are also caught by anglers in various rivers such as the Usk and Wye. **Conservation Status** Endangered.

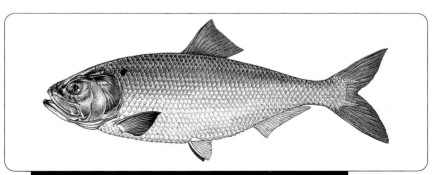

Macedonian Shad *Alosa macedonica*

Size 20–25 cm; maximum 35 cm. **Distinctive features** teeth on palatine and vomer conspicuous; 42–55 lateral scales; 48–50 vertebrae. **Distribution** found only in Lake Volvi in Greece. A purely lentic, freshwater species which does not migrate into streams. Found shoaling in the pelagic zone in summer but goes deeper in winter. **Reproduction** July–August spawning in shallow water, in schools, over gravel and coarse sand. Reaches 10 cm after 1 year; 15–22 cm (23–43 g) after 2 years and 20–25 cm after 4 years. Mature at 1+. **Food** mainly plankton (copepods and cladocerans), but large specimens eat some fish. **Value** caught mainly as a bycatch with other fish; catches variable – 397 tonnes in 1964 but only 1 tonne in 1967. **Conservation Status** Lower Risk.

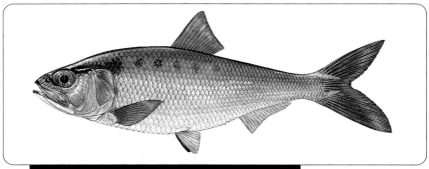

Caspian Shad *Caspialosa caspia*

Size 12–20 cm; maximum 32 cm; mean weight about 100 g. **Distinctive features** upper jaw with median notch, but teeth small or absent; more than 90 gill rakers. *Caspialosa caspia vistonica* Economidis & Sinis 1986 is an endemic subspecies living only in Lake Vistonis. **Distribution** Black and Caspian Seas, and Sea of Azov. Migrates into large rivers (e.g. Danube, Don) in spring. **Reproduction** May–June. Spawns in the lower reaches of rivers and their estuaries. The semipelagic eggs hatch after 2–3 days, and the young move to the sea, maturing in 2–3 years. 400,000–700,000 eggs per female. **Food** mainly invertebrates, especially large zooplankton, and small fish. **Value** an important commercial species which is netted in large numbers in the Caspian Sea. **Conservation Status** Lower Risk.

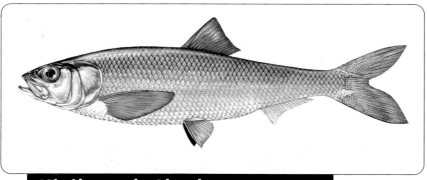

Kizilagach Shad *Caspialosa curensis*

Size 14–16 cm; maximum 19 cm. **Distinctive features** no notch on upper jaw; distinct teeth and large eyes. **Distribution** the Caspian Sea, especially Kizilagach Bay. **Reproduction** April–May in the sea. Relatively little is known of the biology of this species. The young mature at 3–4 years. **Food** invertebrates, especially crustaceans. **Value** of commercial importance, especially in the Kizilagach area where it is netted in some numbers. **Conservation Status** Lower Risk.

Dolginka Shad *Caspialosa maeotica*

Size 30–45 cm; maximum 49 cm. **Distinctive features** marked median notch on upper jaw; vomerine teeth present; eye small. **Distribution** the Caspian Sea and the Sea of Azov. Found near the mouths of some rivers, and occasionally further upstream in fresh water. **Reproduction** April–June, the young mature after 3 years and live up to about 8 years. **Food** invertebrates when young, invertebrates (especially crustaceans) and fishes (e.g. gobies) when adult. **Value** of considerable commercial importance, large numbers being netted in various parts of the Caspian Sea. **Conservation Status** Lower Risk.

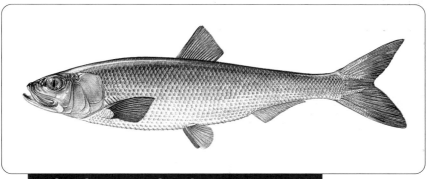

Black Sea Shad *Caspialosa pontica*

Size 15–20 cm; maximum 41 cm; average weight about 69 g. **Distinctive features** marked median notch in upper jaw; 60–90 gill rakers. **Distribution** Black and Caspian Seas and the Sea of Azov. Enters large rivers (e.g. Dneiper, Danube, Volga) in spring and migrates upstream. **Reproduction** April–July, in the middle reaches of rivers, spawning over sand and gravel. Fry migrate downstream in a few months and mature in the sea at 2–3 years. They live for a maximum of 7 years. **Food** invertebrates when young, invertebrates and small fish when adult. **Value** an important commercial species. **Conservation Status** Lower Risk.

Bigeye Shad *Caspialosa saposhnikovi*

Size 20–30 cm; maximum 36 cm; fish with a mean length of 18 cm weigh 84 g. **Distinctive features** upper jaw with marked notch; teeth strong; eyes large. **Distribution** mainly the northern Caspian Sea and the delta of the River Volga, although it never enters the river proper. **Reproduction** April, in estuarine waters in the north-eastern part of the sea. The young mature after 3–4 years and live 6–7 years. **Food** invertebrates, especially crustaceans. **Value** netted commercially (e.g. off the coast of Daghestan), but not as important as other Caspialosa species. **Conservation Status** Lower Risk.

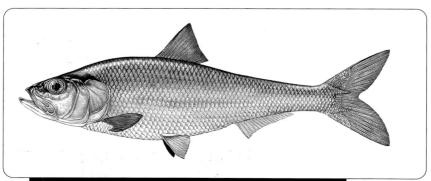

Suworow's Shad *Caspialosa suworowi*

Size 15–20 cm; maximum 27 cm. **Distinctive features** median notch in upper jaw; vomerine teeth present; gill rakers usually 50–55. **Distribution** the Caspian Sea, particularly the northern areas; regularly enters the delta of the Volga. **Reproduction** May. Spawning takes place in the sea, off the Volga delta. **Food** mainly invertebrates. **Value** netted commercially in considerable numbers in the northern Caspian Sea. **Conservation Status** Lower Risk.

Abrau Sardelle *Clupeonella abrau*

Size 6–7 cm; maximum 8.5 cm. **Distinctive features** no notch in upper jaw; less than 53 gill rakers. **Distribution** found only in fresh water; Lake Abrau near Novorossiisk and Abulional Lake near Bursa (Asia Minor). **Reproduction** June–October, the pelagic eggs hatching in about 12 hours. Adults mature after 2 years. **Food** invertebrates, especially opossum shrimps. **Value** of little real commercial or sporting significance. **Conservation Status** Vulnerable.

Tyulka Sardella *Clupeonella cultriventris*

Size 9–12 cm; maximum 17 cm; mean weight about 5 g. **Distinctive features** no notch in upper jaw; abdominal keel; more than 52 gill rakers. **Distribution** Black and Caspian Seas and the lower reaches of their large rivers (e.g. Danube, Bug). One isolated population in Chorkhal Lake (Ural basin). **Reproduction** April–May, spawning in the lower reaches of rivers and in estuaries. The semipelagic eggs hatch in 3–4 days, the young moving to the sea where eggs and larvae are occasionally found. Adults mature at 2–3 years and live up to 5 years. About 30,000 eggs per female. **Food** mainly invertebrates. **Value** an important commercial species, netted in the sea. Used for canning. **Conservation Status** Lower Risk.

Carp Family
Cyprinidae

The Cyprinidae, often known as carp or minnows, is a very large and variable family found in most parts of the world except South America and Australia (though some have been introduced and are now established in these continents). They are all freshwater species and only a few are able to venture occasionally into the brackish water of estuaries. This is the largest family of fish in the world with some 275 genera and about 2,000 species. It is the dominant family in European fresh waters in terms of numbers of species.

Members of the family may be small to large fish of quite variable shape. The protrusible mouth is variable in size and position but never possesses teeth. These are replaced functionally by pharyngeal bones with 1–3 rows of teeth which grind food against a pair of horny pads on the opposite side of the pharynx. These teeth are important in identification, the number of rows, and teeth in each row, being indicated thus: 4.1.1 – which means that the pharyngeal bone concerned has 3 rows of teeth, with 4 teeth on the inside row and 1 on each of the other rows (see fig. 2b). One or two sensory barbels are often present just beside the mouth. Another aid to their sensory perception is the increased hearing provided by a set of three small bones which transmit vibrations to the inner ear from the anterior part of the swimbladder, which acts as a resonating chamber. The fins are all moderately developed and usually soft rayed, except for the stiffened leading rays of the dorsal and anal fins in a few species. The body is usually well covered with cycloid scales and in most fish a sensory lateral line is present.

The sexes often appear different during the spawning season, the males of some species becoming brightly coloured with well-developed tubercles on the head, body and fins; the shape, position, size and number of these tubercles (which in some species also appear on the females) are useful aids in identification. Many species have communal spawning, and hybridisation between several of the species is common – the hybrids usually being intermediate between the parents, a point to remember when trying to identify apparently unusual species. Many of the genera need further study to establish the true status of species and subspecies within them – the fact that leading taxonomists cannot agree even as to the genus which some species in southern Europe should be assigned to is indicative of the present uncertainty in this area.

As might be expected, from the large number of species, the Cyprinidae show considerable differences in habit and habitat, and occupy a variety of niches in fresh waters. Most species feed on invertebrates and only occasionally on fish. A number are omnivorous, feeding both on invertebrates and plants. A few species are purely plant feeders and such fish have proved to be important to humans in two ways – as efficient producers of fish flesh, and as potential controllers of vegetation.

As well as long standing and well established species introduced long ago (e.g. *Cyprinus carpio*, *Carassius auratus*) several other species have been introduced in recent years, and are now well established as part of the European fauna (e.g. *Pimephales promelas*, *Pseudorasbora parva*). However, the status of others introduced in the past is uncertain (e.g. *Mylopharyngodon piceus* (Richardson 1846), *Parabramis pekinensis* (Basilewsky 1855)), and such species are not included in the present checklist.

There are 32 genera and about 145 species in Europe.

Key to genera of Cyprinidae in Europe

1	More than 14 branched rays in dorsal fin; a serrated ray present in dorsal, and usually anal fin	*Go to* **2**
	Less than 14 branched rays in dorsal fin; a serrated ray never present in anal fin	*Go to* **3**
2	Four barbels; pharyngeal teeth triserial	***Cyprinus*** p112
	No barbels; pharyngeal teeth uniserial	***Carassius*** p102
3	Leading ray of dorsal fin distinct from, and much shorter than, 2nd; lips thick	***Pachychilon*** p129
	Leading ray of dorsal fin close to, and about the same size as, 2nd; lips normal	*Go to* **4**
4	Barbels present	*Go to* **5**
	Barbels absent	*Go to* **8**
5	Undulating lateral line; scales absent	***Aulopyge*** p94
	Straight or only moderately curved lateral line; scales present	*Go to* **6**
6	Two pairs of barbels; pharyngeal teeth triserial	***Barbus*** p95
	One pair of barbels; pharyngeal teeth uni- or biserial	*Go to* **7**
7	Mouth terminal; scales very small, more than 80 in lateral line; pharyngeal teeth uniserial	***Tinca*** p150
	Mouth inferior; scales large, less than 50 in lateral line; pharyngeal teeth biserial	***Gobio*** p114
8	Undulating lateral line; abdomen with a scaleless dermal keel starting at throat	***Pelecus*** p130
	Lateral line not undulating; keel, if present, starting behind ventral fins	*Go to* **9**
9	Mouth inferior, a transverse cleft; lower jaw trenchant and covered with cartilage	***Chondrostoma*** p105
	Mouth terminal, oblique or, if inferior, crescentic; never a transverse cleft covered with cartilage	*Go to* **10**
10	Scaleless abdominal keel before vent	*Go to* **11**
	No scaleless keel on abdomen before vent	*Go to* **16**
11	Scaled dorsal keel behind dorsal fin; pharyngeal teeth uniserial	***Vimba*** p152
	No keel behind dorsal fin	*Go to* **12**
12	Pharyngeal teeth uniserial or biserial; scaleless keel behind vent	*Go to* **13**
	Pharyngeal teeth biserial; no scaleless keel behind vent	***Chalcalburnus*** p104
13	Scaleless groove before dorsal fin; scales thick, embedded; more than 20 branched anal rays.	***Abramis*** p89
	No scaleless groove before dorsal fin; scales thin, easily detached; less than 20 branched anal rays	*Go to* **14**
14	Less than 10 branched anal rays; premaxilla with long ascending process; small knob on dentary bone	***Tropidophoxinellus*** p151
	More than 10 branched anal rays; no long ascending process on premaxilla; large knob on dentary	*Go to* **15**
15	Gill rakers long, crowded; pharyngeal teeth usually serrated	***Alburnus*** p92
	Gill rakers short, well spaced; pharyngeal teeth not serrated	***Alburnoides*** p91
16	Lateral line incomplete, short, ending well before dorsal fin	*Go to* **17**
	Lateral line complete, or if incomplete, ending below or behind dorsal fin	*Go to* **20**

Key to genera of Cyprinidae in Europe *continued*

17 Lateral stripe dark, running full length of body; 4–14 scales along lateral line	*Go to* **18**
Lateral stripe pale silver-blue or green-blue	*Go to* **19**
18 Ventral fin with 8 rays; lateral line short, only 4–6 scales long	**Pseudophoxinus** p137
Ventral fin with 9–10 rays; more than 6 scales on lateral line	**Anaecypris** p93
19 Mouth subterminal; body deep; 8–10 branches rays in anal fin	**Rhodeus** p140
Mouth terminal; body elongate; 10–13 branches rays in anal fin; lower jaw with a tubercle entering a notch in upper jaw	**Leucaspius** p119
20 Scales very thin, rudimentary or absent	**Phoxinellus** p131
Scales normal	*Go to* **21**
21 Branchiostegal membranes attached under eye; lower jaw with a tubercle entering notch in upper jaw; mouth terminal, very large	**Aspius** p94
Branchiostegal membranes attached behind eye; upper jaw not notched to receive tubercle in lower jaw; mouth inferior or subterminal, if terminal then small	*Go to* **22**
22 Lateral line ending below dorsal fin	*Go to* **23**
Lateral line complete	*Go to* **24**
23 More than 40 lateral scales; male with fleshy dorsal pad behind head	**Pimephales** p136
Less than 40 lateral scales; males without pad behind head	**Ladigesocypris** p119
24 Origin of dorsal fin slightly behind vertical from posterior end of ventral fin base; pharyngeal teeth biserial	*Go to* **25**
Origin of dorsal fin above ventral fins	*Go to* **29**
25 Scales large, less than 45 along lateral line	**Scardinius** p148
Scales small, more than 45 along lateral line	*Go to* **26**
26 Mouth inferior; less than 60 lateral scales	**Iberocypris** p118
Mouth superior; more than 60 lateral scales	*Go to* **27**
27 Eyes below head midline; more than 100 lateral scales	**Hypophthalmichthys** p117
Eyes above head midline; less than 100 lateral scales	*Go to* **28**
28 Body deep, maximum depth exceeding length of caudal peduncle, and more than 24 per cent of body length	**Eupallasella** p113
Body elongate, maximum depth less than length of caudal peduncle, and less than 24 per cent of body length	**Phoxinus** p134
29 Eyes small, on side of head near midline	**Ctenopharyngodon** p111
Eyes large, on side of head above midline	*Go to* **30**
30 Pharyngeal teeth uniserial	*Go to* **31**
Pharyngeal teeth biserial; usually 8 or less branched rays in dorsal fin	**Leuciscus** p120
31 Mouth superior; 7 branched rays in dorsal fin; lateral line straight	**Pseudorasbora** p139
Mouth terminal or inferior; more than 8 branched rays in dorsal fin; lateral line with bend	**Rutilus** p141

Abramis:

Four species of this genus are found in Europe and may be separated as follows:

1	Pharyngeal teeth uniserial; anal fin with 23–43 branched rays	*Go to* **2**
	Pharyngeal teeth biserial	***Abramis bjoerkna*** p90
2	Less than 30 branched rays in anal fin; 51–60 scales along lateral line; mouth subterminal	***Abramis brama*** p90
	More than 30 branched rays in anal fin	*Go to* **3**
3	Lateral line with 49–52 scales; mouth subterminal	***Abramis sapa*** p91
	Lateral line with 65–75 scales; mouth terminal	***Abramis ballerus*** p89

Blue Bream *Abramis ballerus*

Size 20–30 cm; maximum 45 cm; specimens with an average length of 18 cm weigh about 92 g. **Distinctive features** deep, laterally compressed body with 66–73 rather small scales along lateral line; dorsal fin with 8–9 and anal fin with 35–44 branched rays. **Distribution** found in slow-flowing rivers and lakes in lowland areas of central and eastern Europe whose basins enter the Baltic, Black and Caspian Seas, in the brackish waters of which it also occurs. **Reproduction** April–June among plants in shallow water. 4,200–25,400 eggs per female. Mature at 3–4 years. **Food** invertebrates, especially planktonic crustaceans. **Value** of little commercial or sporting value. **Conservation Status** Lower Risk.

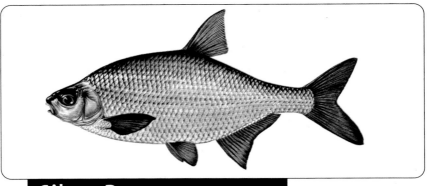

Silver Bream *Abramis bjoerkna*

Size 20–30 cm; maximum 35 cm; maximum weight 1.25 kg; British rod record 425 g (1988). **Distinctive features** body deep and strongly compressed; 40–45 scales along lateral line; dorsal fin with 8–9 and anal fin with 19–24 branched rays. Formerly *Blicca bjoerkna* **Distribution** found in large slow-flowing rivers and in lakes, in much of central and northern Europe from eastern England to the Caspian Sea. **Reproduction** May–July, communal spawning among plants in shallow water. The young mature after 3–5 years and may live up to 10 years; 11,000–82,000 eggs per female. Hybridises with *Abramis brama*, *Rutilus*, *Scardinius* and *Vimba*. **Food** zooplankton when young; when older, invertebrates, especially worms, molluscs and insect larvae. **Value** of some local importance both commercially (in net and trap fisheries) and as a sport species. **Conservation Status** Lower Risk.

Common Bream *Abramis brama*

Size 30–50 cm; maximum 80 cm; British rod record 7.512 kg (1991). **Distinctive features** body very deep and compressed laterally; 51–60 scales along lateral line; dorsal fin with 8–10 and anal fin with 24–30 branched rays. **Distribution** found in slow-flowing rivers and lakes throughout much of Europe from Ireland to the Aral Sea. Introduced to Ireland. It also occurs in estuarine and brackish water in some areas (e.g. Gulf of Finland). **Reproduction** May–July, communal spawning at night among weeds in shallow water. The eggs hatch in 5–10 days and the young mature after 3–5 years. 104,000–587,000 eggs per female. Hybridises with *Abramis bjoerkna*, *Rutilus* and *Scardinius*. **Food** uses an extendible tubular mouth to feed on benthic invertebrates, especially worms, molluscs and insect larvae. **Value** of commercial value in central Europe where it is caught in nets and traps. Also important as a sport species in many countries. **Conservation Status** Lower Risk.

Whiteye Bream *Abramis sapa*

Size 15–25 cm; maximum 39 cm; maximum weight 800 g. **Distinctive features** mouth subterminal; deep, laterally compressed body with 48–53 scales along lateral line; dorsal fin with 8 and anal fin with 35–42 branched rays. **Distribution** found in slow-flowing rivers, some lakes, and in brackish water in several rivers basins (e.g. Danube and Volga) entering the northern Black and Caspian Seas. Sea populations migrate to fresh water to spawn. **Reproduction** April–May communally among weed in running water. The young mature after 3–4 years and may live for more than 8 years. 8,000–150,000 eggs per female. **Food** when young on zooplankton, when older on benthic invertebrates, mainly molluscs, crustaceans and insect larvae. **Value** of some local commercial importance in net fisheries and sometimes sought as a sport fish. **Conservation Status** Lower Risk.

Alburnoides: only one species occurs in Europe:

Schneider *Alburnoides bipunctatus*

Size 10–12 cm; maximum 16 cm. **Distinctive features** terminal mouth; pharyngeal teeth smooth; gill rakers short and set wide apart; 44–52 scales along lateral line, which is dark; dorsal fin with 7–9 and anal fin with 11–17 branched rays. **Distribution** found in streams, rivers and occasionally lakes in parts of Europe from western France east to beyond the Caspian Sea. **Reproduction** May–July, spawning over gravel and small stones in running water. **Food** invertebrates, mainly insect larvae and adults; some surface feeding. **Value** of no commercial value, but occasionally used as a bait species or kept in aquaria. **Conservation Status** Lower Risk.

Alburnus: Three species are found in Europe and may be separated as follows:

1	Body slim, maximum depth about twice that of caudal peduncle	*Go to* **2**
	Body stout, maximum depth about three times that of caudal peduncle	***Alburnus charusini*** p93
2	More than 17 rays in anal fin	***Alburnus alburnus*** p92
	Less than 18 rays in anal fin	***Alburnus albidus*** p92

White Bleak *Alburnus albidus*

Size 10–15 cm; maximum 20 cm. **Distinctive features** body slim, maximum depth about twice that of caudal peduncle; 42–51 lateral scales; 13–18 branched rays in anal fin. **Distribution** found in streams, rivers and some lakes only in waters in the former Yugoslavia and in north and south Italy. **Reproduction** June–August, spawning in shoals over gravel and weed. **Food** invertebrates, especially worms, crustaceans, insect larvae and adults; zooplankton when young; sometimes surface feeding. **Value** of little commercial or sporting value. **Conservation Status** Endangered.

Common Bleak *Alburnus alburnus*

Size 12–15 cm; maximum 25 cm; British rod record 120 g (1982). **Distinctive features** body slim; scaleless keel from bases of ventral fins to anus; 48–45 lateral scales; dorsal fin with 8–9 and anal fin with 16–20 branched rays; base of anal fin longer than that of dorsal fin; more than 17 rays in anal fin. **Distribution** found in slow-flowing rivers and lakes throughout much of Europe from France east to the Caspian Sea; occasionally enters brackish water. **Reproduction** May–July among stones and gravel in shallow water. The eggs hatch in 5–10 days and the young mature after 2–3 years; may live for 6 years. 5,000–6,500 eggs per female. Hybridises with *Leuciscus*, *Rutilus* and *Scardinius*. **Food** invertebrates, especially crustaceans and insects (larvae and adults), often insects on the surface. **Value** formerly important commercially locally, when it was used for animal food or to produce a material for coating artificial pearls from the fishes' scales. Of some sporting value. **Conservation Status** Lower Risk.

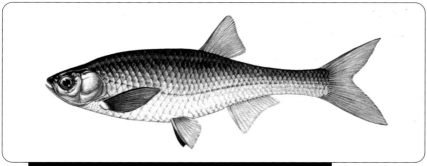

Caucasian Bleak *Alburnus charusini*

Size 8–10 cm; maximum 12 cm. **Distinctive features** maximum depth of body about 3 times that of caudal peduncle; 41–47 scales along lateral line; dorsal fin with 7–9 and anal fin with 14–17 branched rays. **Distribution** found in lakes and slow-flowing rivers in several river basins (e.g. Kuma, Terek, Sulak) in Transcaucasia between the Black and Caspian Seas. **Reproduction** little is known about the reproductive habits of this species. **Food** invertebrates, mainly crustaceans, insect larvae and adults. **Value** of no commercial or sporting significance. **Conservation Status** Lower Risk.

Anaecypris: only one species of this genus occurs in Europe:

Spanish Minnowcarp *Anaecypris hispanica*

Size 4–6 cm; maximum 8 cm. **Distinctive features** superior mouth; scaleless keel from pelvic fins to anus; dark stripe along each side; lateral line extends for only 6–14 scales; 59–71 lateral scales; dorsal fin with 6–7 and anal fin with 8–10 branched rays. **Distribution** found only in small vegetated streams in the River Guadiana and Guadalquivir basins in Portugal and southern Spain. **Reproduction** shoaling in small streams during May–July. **Food** mainly invertebrates, especially insects; some plant material (algae). **Value** of no commercial or sporting value. **Conservation Status** Endangered.

Aspius: only one species of this genus is found in Europe:

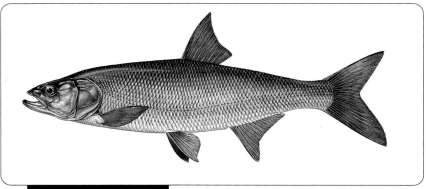

Asp *Aspius aspius*

Size 50–75 cm; maximum 100 cm, exceptionally 120 cm; maximum weight about 12 kg. **Distinctive features** terminal mouth large, lower jaw with a tubercle entering notch in upper jaw; body with well-developed scales, 64–76 along lateral line; keel between pectorals and vent. **Distribution** found in the middle reaches of rivers in central Europe from The Netherlands to the west of the Caspian Sea. Occurs occasionally in lakes and in brackish water in estuarine areas. **Reproduction** April–June, spawning among stones and gravel in flowing water. The eggs hatch in 10–15 days and the adults mature at 4–5 years of age. 58,000–500,000 eggs per female. **Food** invertebrates (especially crustaceans) when young and shoaling, mainly fish when larger and a solitary predator. **Value** of considerable commercial value in some areas where it is caught in nets and traps (e.g. Volga). It is also prized as a sport fish in some areas. **Conservation Status** Vulnerable.

Aulopyge: only one species of this genus occurs in Europe:

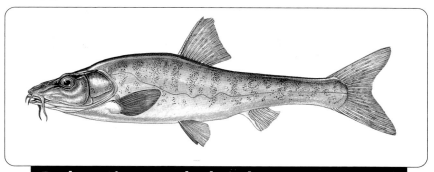

Dalmatian Barbelgudgeon *Aulopyge hugeli*

Size 9–12 cm; maximum 13 cm. **Distinctive features** a well-developed snout with 4 barbels; body scaleless, but with an undulating lateral line running from head to caudal fin. **Distribution** found only in running water in parts of Croatia and Bosnia, including subterranean reaches of karst rivers. **Reproduction** habits little known. **Food** bottom invertebrates, especially worms and insect larvae. **Value** of minor local commercial significance, but no sporting value. **Conservation Status** Critically Endangered.

Barbus: this is a very large genus with considerable uncertainty remaining over the status of the various species and subspecies. 23 species are recognised in the species list for Europe but only 11 of these are discussed fully below. Considerable work remains to be done on the taxonomy of those which are not dealt with in full here (i.e. *Barbus caninus, Barbus euboicus, Barbus graellsii, Barbus guiraonis, Barbus haasi, Barbus macedonicus, Barbus microcephalus, Barbus peloponnesius, Barbus plebejus, Barbus sclateri, Barbus steindachneri, Barbus tyberinus*); those covered in full may be identified as follows:

1	Longest dorsal fin ray stiff, and with a serrated hind edge for most of its length	Go to **2**
	Longest dorsal fin ray without serrations, or if present they occur only on the lower half	Go to **3**
2	Lateral line with 55–65 scales	***Barbus barbus*** p96
	Lateral line with 49–51 scales	***Barbus comizo*** p99
3	Body with dark spots or mark	Go to **4**
	Body without dark spots or marks	Go to **7**
4	Numerous distinct small dark spots on body and fins, but not on head; 12 rays in dorsal fin	Go to **5**
	Many large dark marks often forming a mosaic on back and fins, and extending onto head; 11 rays in dorsal fin; 48–55 scales in lateral line	***Barbus meridionalis*** p100
5	Barbels long, more than 3 times eye diameter, 60–71 scales in lateral line	***Barbus ciscaucasicus*** p98
	Barbels short, less than 3 times eye diameter	Go to **6**
6	Lower part of longest dorsal ray slightly serrated; belly yellowish; 49–52 scales in lateral line	***Barbus albanicus*** p96
	Longest dorsal ray without serrations; belly whitish	***Barbus prespensis*** p101
7	Barbels very short, equal to, or less than, eye diameter	***Barbus cyclolepis*** p99
	Barbels longer, greater than eye diameter	Go to **8**
8	Usually more than 60 scales in lateral line	Go to **9**
	Less than 60 scales in lateral line	Go to **10**
9	Usually 7 branched dorsal rays; barbels long, anterior barbel reaching beyond anterior eye margin; 67–76 lateral scales	***Barbus brachycephalus*** p97
	Usually 8 branched dorsal rays; barbels short, anterior barbel never reaching anterior eye margin; 58–65 lateral scales	***Barbus capito*** p98
10	50–60 lateral scales; anterior barbels 75% of length of posterior	***Barbus graecus*** p100
	46–51 lateral scales; anterior barbels 50% of length of posterior	***Barbus bocagei*** p97

Albanian Barbel *Barbus albanicus*

Size 30–40 cm; maximum 45 cm. **Distinctive features** 4 short barbels around inferior mouth, each less than 3 times eye diameter; 49–52 scales along lateral line; lower part of longest dorsal ray slightly serrated posteriorly. **Distribution** found only in ponds and rivers in parts of Greece. **Reproduction** May–July, spawning over gravel and stones. **Food** mainly benthic invertebrates, especially insect larvae. **Value** of little commercial but some sporting value. **Conservation Status** Lower Risk.

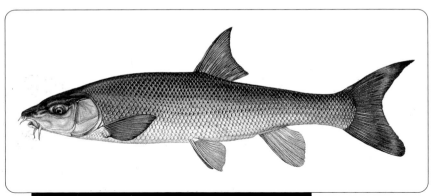

Common Barbel *Barbus barbus*

Size 25–75 cm; maximum 100 cm; maximum weight 10 kg; British rod record 7.314 kg (1994). **Distinctive features** inferior mouth with 4 fleshy sensory barbels; 55–65 lateral scales; last unbranched ray in dorsal fin thickened and with numerous denticles posteriorly; dorsal fin with 7–8 and anal fin with 5 branched rays. **Distribution** found in the middle reaches of rivers in middle Europe from eastern England to the Black Sea. **Reproduction** May–July among gravel and stones in flowing water. The yellow adhesive eggs (which are reputed to be poisonous) hatch in 10–15 days and the young mature after 4–5 years. 3,000–32,000 eggs per female. **Food** active at night, eating bottom invertebrates (mainly worms, molluscs and insect larvae) and some plant material when young; invertebrates and small fish when older. **Value** of some commercial importance locally, where it is caught by traps and nets. An important angling species in several countries. **Conservation Status** Lower Risk.

Boca Barbel *Barbus bocagei*

Size 40–50 cm; maximum 60 cm. **Distinctive features** snout prominent; anterior barbels short and thin, posterior barbels long; scales large and obvious, 46–51 scales along lateral line. **Distribution** occurs in clean rivers over much of Spain and Portugal. **Reproduction** May–June, spawning over areas of sand and gravel. **Food** mainly benthic invertebrates, especially crustaceans, molluscs and insect larvae; sometimes on plant material. **Value** important locally to anglers. **Conservation Status** Lower Risk.

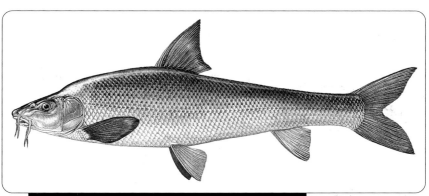

Aral Barbel *Barbus brachycephalus*

Size 80–100 cm; maximum 120 cm; maximum weight about 22.5 kg. **Distinctive features** 4 barbels around inferior mouth reaching back beyond margin of eye; 67–76 lateral scales; dorsal fin with 6–8 and anal fin with 5–6 branched rays. **Distribution** brackish waters of the Aral and Caspian Seas, and in fresh water only in the lower reaches of rivers there (e.g. Volga, Ural), often only for spawning. **Reproduction** April–July, migrating upstream in rivers easily accessible from the sea. Up to 1,259,000 eggs per female. **Food** bottom invertebrates (mainly molluscs in the Aral Sea). **Value** of commercial importance in the Aral and Caspian areas, where fish are caught in nets and traps during the spawning migration. **Conservation Status** Lower Risk.

Dog Barbel *Barbus caninus* recorded from Italy (type locality: Po basin) and Switzerland (Lago Maggiore). **Conservation Status** Endangered.

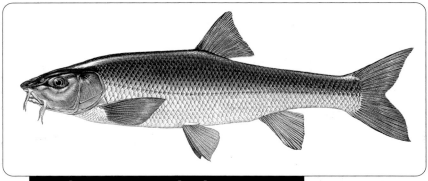

Bulatmai Barbel *Barbus capito*

Size 60–90 cm; maximum 105 cm. **Distinctive features** 4 barbels around inferior mouth, anterior barbel never reaching margin of eye; 58–65 lateral scales; dorsal fin with 8 and anal fin with 5 branched rays. **Distribution** lower and middle reaches of rivers entering Caspian and Aral Seas; also in salt water in these areas, migrating into fresh water at the spawning season. **Reproduction** in running water, young maturing in 6–7 years. **Food** benthic invertebrates, especially insect larvae. Bottom invertebrates and plant material when young; and fish when older. **Value** of local commercial value, taken in traps and nets in rivers during the spawning season. **Conservation Status** Lower Risk.

Caucasian Barbel *Barbus ciscaucasicus*

Size 25–35 cm; maximum 39 cm. **Distinctive features** 4 long barbels around inferior mouth, these being more than 3 times the diameter of the eye in length; 60–71 scales along lateral line; dorsal fin with 8 and anal fin with 5 branched rays; body, especially above the lateral line, covered with numerous dark spots. **Distribution** found only in the middle and upper reaches of rivers entering the western Caspian Sea (e.g. Kuma and Terek); ascends quite high into mountainous regions. **Reproduction** May–July, spawning over stones and gravel, sometimes in quite small tributaries. **Food** mainly invertebrates, especially insect larvae. **Value** of minor commercial, and no angling importance. **Conservation Status** Lower Risk.

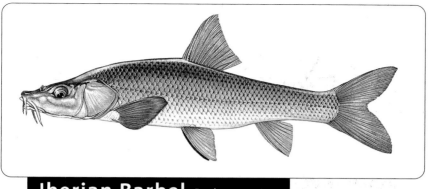

Iberian Barbel *Barbus comizo*

Size 20–30 cm; maximum 35 cm. **Distinctive features** prominent snout with 2 pairs of barbels around inferior mouth; largest dorsal ray stiff with a serrated hind edge; 49–51 lateral scales. **Distribution** in the middle reaches of a few rivers in south-western Portugal and Spain (e.g. Guadalquivir, Tagus). **Reproduction** spawns over gravel and stones. **Food** mainly benthic invertebrates, especially molluscs and insect larvae, but also some plant material when young and fish when older. **Value** of no commercial but some sporting value. **Conservation Status** Vulnerable.

Turkish Barbel *Barbus cyclolepis*

Size 20–30 cm; maximum 35 cm. **Distinctive features** inferior mouth with 4 short barbels – these are equal to, or less than, eye diameter; third spinous ray of dorsal fin serrated. **Distribution** found among stones and boulders only in running water in a few basins in Greece and Turkey emptying into the Black and Aegean Seas (e.g. Rivers Maritza and Struma). **Reproduction** April–July, over gravel and stones in running water. **Food** mainly invertebrates, but some plant food when young, fish and fish eggs when older. **Value** of no commercial or sporting value. **Conservation Status** Vulnerable.

Euboean Barbel *Barbus euboicus*
recorded from the island of Euboea in Greece (type locality: small streams, Evia).
Conservation Status Endangered.

Carp Family Cyprinidae

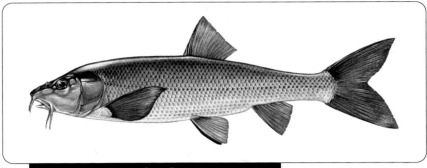

Greek Barbel *Barbus graecus*

Size 30–40 cm; maximum 45 cm. **Distinctive features** body without dark spots or marks; 4 barbels around inferior mouth, each longer than eye diameter; 50–60 scales along lateral line; dorsal fin with 11 branched rays. **Distribution** found only where the current is moderate or slow in a few waters in Albania and Greece (e.g. in the Sperchios River and Lakes Yliki and Paralimni). **Reproduction** habits not known. **Food** benthic invertebrates, especially insect larvae and molluscs; some plant material. **Value** of some commercial but little sporting value. **Conservation Status** Lower Risk.

Graells' Barbel *Barbus graellsii*
recorded from Spain (type locality: River Ebro). **Conservation Status** Lower Risk.

Guiraonis Barbel *Barbus guiraonis*
recorded from Spain (type locality: River Jucar). **Conservation Status** Vulnerable.

Orange Barbel *Barbus haasi*
recorded from Spain (type locality: Noguera Pallaresa stream).
Conservation Status Vulnerable.

Macedonian Barbel *Barbus macedonicus*
recorded from Greece and Macedonia (type locality: River Vardar).
Conservation Status Vulnerable.

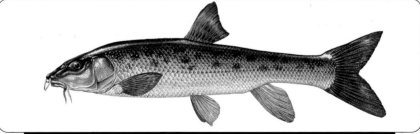

Mediterranean Barbel *Barbus meridionalis*

Size 20–30 cm; maximum 40 cm; a 20 cm fish weighs about 150 g. **Distinctive features** two pairs of barbels, the second pair long and slender; 48–55 lateral scales; longest ray of dorsal fin without posterior serrations; numerous large dark spots, often forming a mosaic on back and fins. **Distribution** bottom living in the middle reaches of rivers in south-western and central Europe. **Reproduction** March–June, after an upstream migration, in running water over gravel and stones – sometimes in areas of vegetation. **Food** mainly benthic invertebrates, but some plant food when young, and fish when adult. **Value** of local commercial and sporting significance. **Conservation Status** Lower Risk.

Smallhead Barbel *Barbus microcephalus*
recorded from Portugal (type locality: River Guadiana) and Spain.
Conservation Status Vulnerable.

Peloponnesian Barbel *Barbus peloponnesius*
recorded from Greece (type locality: Moree). **Conservation Status** Lower Risk.

Adriatic Barbel *Barbus plebejus*
recorded from Croatia, Italy (type locality: Lake Como) and Slovenia.
Conservation Status Lower Risk.

Prespa Barbel *Barbus prespensis*

Size 20–25 cm; maximum 30 cm. **Distinctive features** 4 barbels round inferior mouth; longest dorsal ray without denticles posteriorly. **Distribution** found only in the Prespa Lakes (Albania, Greece and Macedonia) and associated waters. **Reproduction** April–July, spawning over gravel and stones. A short-lived species. **Food** mainly bottom invertebrates, especially insect larvae. **Value** of some commercial but little sporting value. **Conservation Status** Vulnerable.

Sclater Barbel *Barbus sclateri*
recorded from Portugal and Spain (type locality: River Guadaquivir).
Conservation Status Lower Risk.

Steindachner's Barbel *Barbus steindachneri*
recorded from Portugal (type locality: River Guadiana). **Conservation Status** Lower Risk.

Tyber Barbel *Barbus tyberinus*
recorded from Italy (type locality: River Tevere). **Conservation Status** Lower Risk.

Carassius: a small genus with only two species in Europe, though some regard the subspecies *Carassius auratus gibelio* (Bloch 1782) as a third full species. The two full species may be separated as follows:

1 Less than 31 scales along lateral line; more than 34 gill rakes on 1st gill arch; dorsal fin concave; 1st dorsal ray strong, coarsely serrated	***Carassius auratus*** p102
More than 31 scales along lateral line; less than 34 gill rakers on 1st gill arch; dorsal fin convex; 1st dorsal ray feeble, weakly serrated	***Carassius carassius*** p103

normal form

golden form

Goldfish *Carassius auratus*

Size 15–35 cm; maximum 45 cm; maximum weight about 3 kg. **Distinctive features** body of moderate depth and laterally compressed; 27–31 scales along lateral line; dorsal fin concave, first ray strong and coarsely serrated, 15–19 branched rays, anal fin with 5–6. The Gibel Carp *Carassius auratus gibelio* is a common subspecies **Distribution** native to eastern Asia, this fish is now found in rich ponds, lake and slow-flowing rivers in many parts of Europe to which it has been introduced. Many introductions also to other parts of the world including the United States and Canada where it is now established in a large number of waters. **Reproduction** May–July, spawning among thick weed in shallow water. The golden eggs hatch in 5–8 days and the young mature at 2–4 years. They may live up to 20 years or more. 160,000–383,000 eggs per female. Hybridises with *Carassius carassius*. **Food** young feed on zooplankton; when older on invertebrates and plant material. **Value** of some considerable commercial importance since the golden variety is one of the most popular aquarium fish, and large numbers are reared on fish farms for this purpose. Of little sporting value – occasionally used as a bait species. **Conservation Status** Alien.

Crucian Carp *Carassius carassius*

Size 20–45 cm; maximum 50 cm; maximum weight 5 kg; British rod record 2.594 kg (1994). **Distinctive features** no barbels; body usually deep and laterally compressed; 31–36 scales along lateral line; dorsal fin convex, first ray feeble and weakly serrated and with 14–21, anal fin with 6–8 branched rays. **Distribution** found in ponds, lakes and slow-flowing rivers throughout all of eastern and central Europe and many parts of the west. It is not native to many of these areas, but has been successfully introduced. Also introduced elsewhere in the world, including the United States, where a population established near Chicago in the early 1900s has since died out. It is very resistant to deoxygenation and freezing. **Reproduction** May–June, spawning in shallow water among thick weed growth. The yellow eggs hatch after 5–8 days and the young mature after 3–4 years. 137,000–244,000 eggs per female. Some populations are all female (gynogenetic). Hybridises with *Cyprinus* and *Gobius*. **Food** bottom-dwelling invertebrates and plants. **Value** of limited commercial importance (mainly in fish farms), but a useful sport species in some countries. **Conservation Status** Alien.

Carp Family Cyprinidae

Chalcalburnus: a small genus with only two species in Europe which may be separated as follows:

1	Found only in the Prespa Lakes (Adriatic Sea drainage)	*Chalcalburnus belvica*
	Found only in the Black and Caspian Sea drainages	*Chalcalburnus chalcoides*

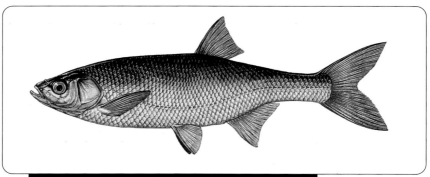

Prespa Bleak *Chalcalburnus belvica*

Size 15–25 cm; maximum 30 cm. **Distinctive features** slim, elongate body; abdomen laterally compressed into a keel, which is free of scales posteriorly. **Distribution** found only in the Prespa Lakes (Greece, Albania and Macedonia) and their associated running waters. **Reproduction** April–August, spawning over gravel and stones in shallow water. **Food** mainly invertebrates, especially worms, molluscs and insect larvae; some plant material. **Value** of some local commercial importance. **Conservation Status** Lower Risk.

Shemaya Bleak *Chalcalburnus chalcoides*

Size 15–30 cm; maximum 40 cm. **Distinctive features** slim, elongate body with 57–70 scales along lateral line; abdomen laterally compressed into a keel, which is free of scales posteriorly; dorsal fin with 8–9 and anal fin with 15–19 branched rays. **Distribution** found in slow flowing parts of rivers and in some lakes in most river basins entering the Black Sea and the western half of the Caspian Sea. It also occurs in salt water in these areas, but migrates into fresh water to breed. **Reproduction** May–July, spawning over gravel and stones in shallow fresh water. 15,500–23,500 eggs per female. Hybridises with *Leuciscus*. **Food** invertebrates, especially worms, molluscs, crustaceans and insect larvae. **Value** of some commercial value in the Caspian area where large numbers are netted during the spawning migration. **Conservation status** Vulnerable.

Chondrostoma: this is a very large genus with considerable uncertainty remaining over the status of the various species and subspecies. 19 species are recognised in the species list for Europe but only 9 of these are discussed fully below. Considerable work remains to be done on the taxonomy of those which are not dealt with in full here (i.e. *Chondrostoma arrigonis, Chondrostoma duriense, Chondrostoma kubanicum, Chondrostoma miegii, Chondrostoma prespense, Chondrostoma scodrense, Chondrostoma turiense, Chondrostoma vardarense, Chondrostoma variabile, Chondrostoma willkommii*); those covered in full may be identified as follows:

1	Less than 56 or more than 66 scales in lateral line	*Go to* **2**
	56–66 scales in lateral line	*Go to* **3**
2	Lateral line with 52–54 scales	***Chondrostoma kneri*** p107
	Lateral line with 88–90 scales	***Chondrostoma phoxinus*** p108
3	Darkish line along side; pharyngeal teeth 5.5 (occasionally 5.6)	***Chondrostoma genei*** p106
	No darkening along side, or if so, pharyngeal teeth 6.5	*Go to* **4**
4	Body deep, greatest depth more than twice that of head	***Chondrostoma soetta*** p109
	Body less deep, greatest depth less than twice that of head	*Go to* **5**
5	Pharyngeal teeth 6.6 or 7.6	*Go to* **6**
	Pharyngeal teeth 6.5 (occasionally 5.5)	*Go to* **7**
6	11 rays in dorsal fin, 12 rays in anal fin; pharyngeal teeth usually 6.6	***Chondrostoma toxostoma*** p110
	12 rays in dorsal fin, 13–14 rays in anal fin; pharyngeal teeth usually 7.6	***Chondrostoma nasus*** p107
7	Snout small and rounded, its length usually less than half eye diameter	***Chondrostoma polylepis*** p109
	Snout prominent, length usually more than half eye diameter	*Go to* **8**
8	Mouth cleft straight; depth of body usually more than 0.22 times length; sometimes with dark lateral band	***Chondrostoma colchicum*** p106
	Mouth cleft rounded; depth of body usually less than 0.22 times length; never with dark lateral band	***Chondrostoma oxyrhynchus*** p108

Arrigon Nase *Chondrostoma arrigonis*
recorded from the Jucar River in eastern Spain (type locality: River Jucar).
Conservation Status vulnerable.

Caucasian Nase *Chondrostoma colchicum*

Size 20–25 cm; maximum 29 cm. **Distinctive features** snout well developed, mouth cleft straight and transverse; lower lip horny and with a sharp edge; 57–64 lateral scales; dorsal fin with 8–9 and anal fin with 9–10 branched rays. Often a longitudinal band of dark spots runs laterally from head to tail. **Distribution** found in the middle reaches of large rivers (e.g. Kuban) in the Caucasian area of southern Russia. **Reproduction** Breeding habits little known. Males with well-developed tubercles. **Food** invertebrates (molluscs and insect larvae) and attached algae and other small plants. **Value** of little commercial, and no sporting value. **Conservation status** Data Deficient.

Durien Nase *Chondrostoma duriense*
recorded from the Jucar River in eastern Spain (type locality: River Jucar).
Conservation Status Vulnerable.

Laska Nase *Chondrostoma genei*

Size 15–20 cm; maximum 30 cm. **Distinctive features** snout and lips moderately developed, lower lip hard and horny; 52–62 lateral scales; dorsal fin with 8 and anal fin with 8–10 branched rays; darkish band along either side. **Distribution** found mainly as small groups in pools in the upper and middle reaches of large rivers in central and northern Italy (e.g. Po). **Reproduction** March–May, spawning in flowing water over stones and gravel. **Food** some invertebrates but mainly fine plant material scraped off rocks. **Value** of little commercial or sporting value. **Conservation Status** Vulnerable.

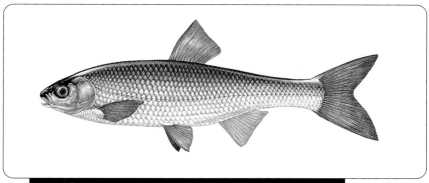

Dalmatian Nase *Chondrostoma knerii*

Size 15–18 cm; maximum 20 cm. **Distinctive features** mouth semicircular in shape; lower lip hard and horny, with sharp edge; scales relatively large, 52–54 along lateral line. **Distribution** found only in the middle reaches of rivers in western Croatia. **Reproduction** March–May, spawning over gravel and stones in running water. **Food** some invertebrates but mainly attached algae scraped off rocks. **Value** of no commercial or sporting value. **Conservation Status** Endangered.

Kuban Nase *Chondrostoma kubanicum*
recorded from the Kuban River in the northwest Caucasus. **Conservation Status** Vulnerable.

Spanish Nase *Chondrostoma miegii*
recorded from several river systems in northeast Spain. **Conservation Status** Vulnerable.

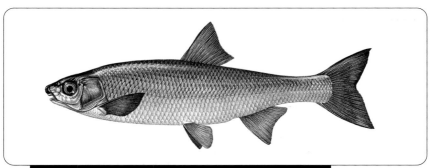

Common Nase *Chondrostoma nasus*

Size 20–40 cm; maximum 51 cm; maximum weight 2.5 kg. **Distinctive features** prominent snout and well-developed lips, the lower lip forming a sharp edged horny pad; mouth inferior; 53–66 lateral scales; dorsal fin with 8–10 and anal fin with 13–14 branched rays. **Distribution** found in the fast flowing sections of the middle reaches of large rivers in central and eastern Europe. **Reproduction** March–May, spawning, after migrating upstream, over gravel and stones in flowing water. The adults mature after 3–4 years and may live for at least 9 years. On average, about 10,000 eggs per female. **Food** mainly material grazed off the surface of stones (e.g. filamentous algae and various invertebrates). **Value** of considerable commercial importance in the Dnieper, Volga and other rivers where it is caught in nets and traps. It is of little sporting significance. **Conservation Status** Lower Risk.

Terek Nase *Chondrostoma oxyrhynchum*

Size 15–20 cm; maximum 24 cm. **Distinctive features** moderate snout, mouth cleft rounded; body elongate, its length about 5 times its depth; 57–66 lateral scales; dorsal fin with 7–8 and anal fin with 9–10 branched rays; never with a dark lateral band. **Distribution** found only in rivers entering the western Caspian Sea (e.g. Kuma and Sulak). **Reproduction** reproductive habits unknown. **Food** invertebrates and especially attached algae which are scraped off rocks, etc. **Value** of little commercial and no sporting value. **Conservation Status** Lower Risk.

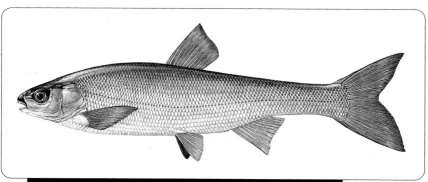

Minnow Nase *Chondrostoma phoxinus*

Size 10–12 cm; maximum 15 cm. **Distinctive features** very small and slim with small scales; 88-90 along lateral line. **Distribution** found only in fast-running water in certain rivers in Croatia and Bosnia. **Reproduction** reproductive habits not known. **Food** some invertebrates but mainly attached algae and other fine plant material scraped off rocks. **Value** of no commercial or sporting value. **Conservation Status** Vulnerable.

Iberian Nase *Chondrostoma polylepis*

Size 15–20 cm; maximum 40 cm. **Distinctive features** usually small, with a rounded snout;. lower lip horny with cutting edge; 59–78 lateral scales; anal fin with 8–10 branched rays. **Distribution** found only in fast flowing stretches of rivers in Portugal (e.g. Douro, Mondego), and western Spain. **Reproduction** reproductive habits unknown. **Food** some benthic invertebrates (especially molluscs and insect larvae) but especially various plants (e.g. attached algae). **Value** of little commercial, and no sporting value. **Conservation Status** Lower Risk.

Prespa Nase *Chondrostoma prespense*
recorded from Greece and Macedonia (type locality: Prespa Lakes). **Conservation Status** Lower Risk.

Skadar Nase *Chondrostoma scodrense*
recorded from Montenegro (type locality: Lake Skadar – but now believed to be extinct). **Conservation Status** Extinct in the Wild.

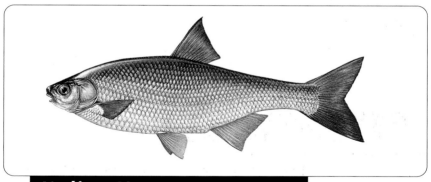

Italian Nase *Chondrostoma soetta*

Size 25–35 cm; maximum 45 cm. **Distinctive features** well-developed lips, the lower lip with a horny cutting edge; body deep, greatest depth more than twice that of head; 55–63 lateral scales; dorsal fin with 8–9 and anal fin with 11–13 branched rays. **Distribution** found only in the middle reaches of large alpine rivers in northern Italy (e.g. Po) and some large lakes there. **Reproduction** April–May, spawning in small groups in shallow water. **Food** some invertebrates (especially molluscs and insect larvae) but mainly fine plant material (algae) scraped off stones. **Value** of some local commercial importance in net and trap fisheries. Not angled for. **Conservation Status** Lower Risk.

Carp Family Cyprinidae

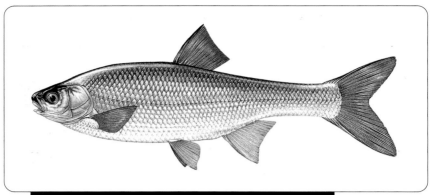

French Nase *Chondrostoma toxostoma*

Size 20–25 cm; maximum 30 cm. **Distinctive features** snout small; mouth small and arched; lower lip hard and horny; 52–56 lateral scales; dorsal fin with 9 and anal fin with 10–11 branched rays; faint dark lateral band. **Distribution** found only in the upper reaches of rivers, and occasionally in montane lakes, in south-western France, Spain and Portugal. **Reproduction** March–May, spawning over gravel and stones in fast-flowing water, often in quite small tributaries. **Food** some invertebrates but mainly fine plant material (attached algae) scraped off rocks. **Value** of no commercial or sporting value. **Conservation Status** Vulnerable.

Turia Nase *Chondrostoma turiense*
recorded from the Turia and Mijares Rivers in northeast Spain. **Conservation Status** Vulnerable.

Vardar Nase *Chondrostoma vardarense*
recorded from Greece and Macedonia (type locality: River Vardar). **Conservation Status** Lower Risk.

Variable Nase *Chondrostoma variabile*
recorded from the Rivers Dnieper, Don, Emba, Ural and Volga in eastern Europe. **Conservation Status** Lower Risk.

Willkomm's Nase *Chondrostoma willkommii*
recorded from Portugal and Spain (type locality: River Guadiana). **Conservation Status** Lower Risk.

Ctenopharyngodon: only 1 species of this introduced genus is currently established in Europe:

Grass Carp *Ctenopharyngodon idella*

Size 40–80 cm; maximum 120 cm; weight 2–5 kg; maximum 30 kg. **Distinctive features** eyes wide apart and low on head; body elongate and cylindrical; mouth inferior; dorsal fin with 7 and anal fin with 7 or 8 branched rays. Lateral scales 43-45. **Distribution** originally from China and the Amur River in Russia. Now widely introduced around the world, including much of Europe and North America. Naturalised in the Danube in Slovakia, Czech Republic, Hungary, Romania and Yugoslavia. Also known as White Amur. **Reproduction** June–August in the main channel of rivers where the pelagic eggs hatch in less than 2 days at 28°C. Growth is rapid – fish can be 50 cm in length after 5 years. They are sexually mature at 6–7 years. 50,000–150,000 eggs per female. **Food** the fry feed initially on zooplankton but soon turn to an almost entirely vegetarian diet, largely of aquatic macrophytes. **Value** of major commercial importance in aquaculture and also used in habitat management to graze down excessive macrophytes. **Conservation Status** Alien.

Cyprinus: a single species of this genus occurs in Europe:

Common Carp *Cyprinus carpio*

Size 25–75 cm; maximum 1.02 m; British rod record 23.358 kg (1980). **Distinctive features** upper lip with 2 long barbels and 2 short barbels; body covered with large scales, 33–40 along lateral line; dorsal fin with 17–22 and anal fin with 5 branched rays. **Distribution** native to eastern Europe and Asia, this fish has been successfully introduced to many other countries all over the world, including the United States and Canada, where it is firmly established. It is found in rich ponds, lakes and slow-flowing rivers. **Reproduction** June–July, spawning among thick weed in shallow water. The eggs hatch in 3–6 days and the young mature after 3–5 years. Very long-lived – 50 years is not uncommon. 93,000–1,664,000 eggs per female. Hybridises with *Carassius*. **Food** bottom-dwelling invertebrates and plant material. **Value** of considerable commercial value in many countries, particularly in central Europe, where many hundreds of fish farms are devoted to this species. Also of notable value as an elusive sport fish, when leather and mirror varieties are often encountered (see fig. 6). The golden and coloured varieties are popular pond fish especially the multi-coloured koi varieties. **Conservation Status** Alien.

Eupallasella: a single species of this genus occurs in Europe:

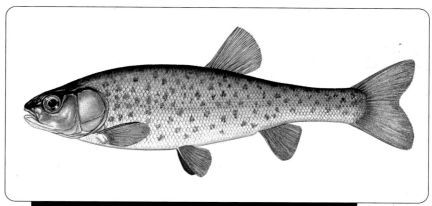

Swamp Minnow *Eupallasella perenurus*

Size 10–15 cm; maximum 19 cm; maximum weight about 100 g. **Distinctive features** scales small, 65–89 along lateral line; dorsal fin with 7 and anal fin with 7–8 branched rays; body deep and covered with small brown spots, but no large blotches. This species used to be placed in the genus *Phoxinus*. **Distribution** found in well-vegetated, rich ponds and lakes in parts of central Europe and in the basins of all rivers which flow into the Arctic Ocean. **Reproduction** June–July, spawning among water plants to which its adhesive eggs are attached. The males do not have nuptial tubercles. The eggs hatch in 5–8 days and the adults mature after 2–3 years. 1,600–18,700 eggs per female. **Food** invertebrates, especially worms, crustaceans and insect larvae; some surface insects. **Value** of considerable commercial importance in parts of Russia, and used as a bait species for sport fishing elsewhere. **Conservation Status** Endangered.

Carp Family Cyprinidae

Gobio: this is another genus where considerable uncertainty still exists over the status of the various species and subspecies. 8 species are recognised in the species list for Europe but only 5 of these are discussed fully below. Considerable work remains to be done on the taxonomy of those which are not dealt with in full here (i.e. *Gobio banarescui*, *Gobio benacensis*, *Gobio elimeius*); those covered in full may be identified as follows:

1 Caudal peduncle short, deep and laterally compressed at origin; its thickness at anal fin equal to, or less than, minimum body depth; its depth at caudal fin more than one-third caudal peduncle length; barbels usually not reaching posterior end of preopercular	***Gobio gobio*** p115
Caudal peduncle long, thick at origin, slender at end; its thickness at anal fin greater than minimum body depth; length of caudal peduncle less than one-third minimum body depth; barbels usually reaching beyond posterior edge of eye	*Go to* **2**
2 Dorsal fin with 8 branched rays; throat naked; colours bright; eye diameter equal to inter-orbital width	***Gobio kessleri*** p116
Dorsal fin with 7 branched rays	*Go to* **3**
3 Throat scaled; eye diameter equal to inter-orbital width	***Gobio uranoscopus*** p116
Throat naked	*Go to* **4**
4 Eye diameter equal to, or slightly greater than, inter-orbital width	***Gobio albipinnatus*** p114
Eye diameter much less than inter-orbital width; colour variegated	***Gobio ciscaucasicus*** p115

Whitefin Gudgeon *Gobio albipinnatus*

Size 10–12 cm; maximum 13 cm. **Distinctive features** 1 pair of long barbels reaching posterior margin of eye; dorsal and caudal fins without dark spots; throat mostly without scales. **Distribution** found only in the slow flowing, deeper waters of the middle and sometimes the lower reaches of rivers in central and eastern Europe, notably the Volga, Dnieper, and Don; also in ponds and lakes on the flood plain. **Reproduction** reproductive habits little known. **Food** bottom invertebrates, mainly worms, crustaceans and insect larvae. **Value** of no commercial or sporting value. **Conservation Status** Vulnerable.

Banarescu's Gudgeon *Gobio banarescui*
recorded from Italy and Macedonia (type locality: River Vardar). **Conservation Status** Vulnerable.

Italian Gudgeon *Gobio benacensis*
recorded from northern Italy (type locality: Lake Garda). **Conservation Status** Lower Risk.

Caucasian Gudgeon *Gobio ciscaucasicus*

Size 11–14 cm; maximum 15 cm. **Distinctive features** elongate snout and inferior mouth; throat usually scaleless; barbels very long, reaching beyond posterior edge of eye; 42–46 scales along lateral line; third spinous ray of dorsal fin serrated. **Distribution** found only in streams and rivers in south-western Russia (Transcaucasia), notably the Kuban, Kuma, Terek, and Sulak. **Reproduction** May–June, but spawning habits little known. **Food** bottom-dwelling invertebrates, especially crustaceans and insect larvae. **Value** of no commercial or sporting significance. **Conservation Status** Lower Risk.

Aliakmon Gudgeon *Gobio elimeius*
recorded from Greece (type locality: Aliakmon River) and Macedonia. **Conservation Status** Vulnerable.

Common Gudgeon *Gobio gobio*

Size 10–15 cm; maximum 20 cm; maximum weight 225 g; British rod record 140 g (1990). **Distinctive features** inferior mouth; 2 well-developed barbels; throat without scales; scales large, 38–45 along lateral line; dorsal fin with 5–7 and anal fin with 6–7 branched rays. **Distribution** a bottom living species found in streams and rivers, and some lakes throughout much of temperate Europe and Asia; introduced to Ireland and Spain. **Reproduction** May–June, among stones and weed in running water. The young hatch after 15–20 days and adults mature after 2–3 years, living up to a maximum of 8 years. 800–3,000 eggs per female. **Food** benthic invertebrates, especially molluscs, crustaceans and insect larvae. **Value** of little commercial significance, but occasionally used as a bait species in sport fishing, or angled for in its own right. **Conservation Status** Lower Risk.

Kessler's Gudgeon *Gobio kessleri*

Size 10–12 cm; maximum 15 cm. **Distinctive features** 1 pair of long barbels reaching hind edge of eye; throat without scales; 40–42 lateral scales; dorsal fin with 8–9 branched rays. **Distribution** found only in fast flowing water in rivers in central Europe in the area of the Danube basin. **Reproduction** reproductive habits not well known. **Food** bottom invertebrates, especially crustaceans and insect larvae. **Value** of no commercial or sporting value. **Conservation Status** Lower Risk.

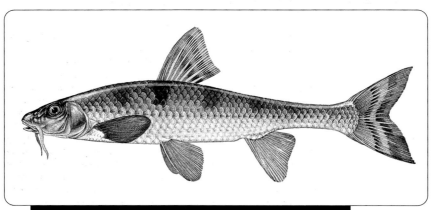

Danube Gudgeon *Gobio uranoscopus*

Size 10–12 cm; maximum 15 cm. **Distinctive features** 1 pair of long barbels which can reach back beyond the eyes; throat with scales; dorsal fin with 7 branched rays; very few spots on body and fins, but a blue band running along lateral line. **Distribution** living in fast-running water in the middle and upper reaches of streams in the Danube basin, in Hungary, Romania, Slovakia and Slovenia. **Reproduction** May–June, among weed and stones in flowing water. **Food** bottom invertebrates, especially worms, crustaceans and insect larvae. **Value** of no commercial or sporting value. **Conservation Status** Endangered.

Hypophthalmichthys: two species of this introduced genus are now
established in Europe. They may be separated as follows:

1 Gill rakers united by a mucous membrane; ventral keel from throat to behind vent; pharyngeal teeth with striated grinding surfaces	*Hypophthalmichthys molitrix* p117
Gill rakers free; ventral keel only behind vent; pharyngeal teeth with smooth grinding surfaces	*Hypophthalmichthys nobilis* p118

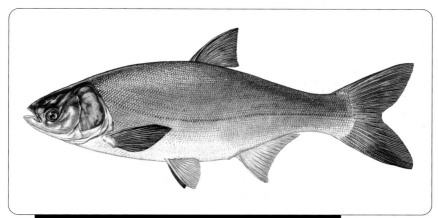

Silver Carp *Hypophthalmichthys molitrix*

Size 80–100 cm; maximum 120 cm; average weight 10 kg. **Distinctive features** deep bodied with an oblique, superior mouth; eye set low – level with the mouth or below; scales small 110–125 along lateral line. Fins dark in colour. **Distribution** native to the River Amur in Siberia where it occurs in slow flowing water and oxbow lakes. Introduced widely to other parts of the world for both aquaculture and the control of algae; it now occurs in many lakes and reservoirs in south east Europe. The same is true of much of the United States, where the species is widespread (though largely still maintained by stocking). **Reproduction** in the Amur basin in May–July in slow flowing water in rivers and lake outlets; most European populations are maintained by the release of stock raised on fish farms. **Food** mainly plankton, especially phytoplankton but some zooplankton, which it filters from the water by means of fine long gill rakers. **Value** of considerable commercial value in parts of Europe and other areas where it has been introduced. **Conservation Status** Alien.

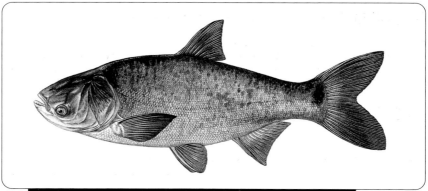

Bighead Carp *Hypophthalmichthys nobilis*

Size 80–100 cm; maximum 120 cm; average weight 10 kg. **Distinctive features** large head, with low set eyes; pharyngeal teeth with smooth grinding surfaces; dark mottled patches on sides; formerly known as *Aristichthys nobilis*. **Distribution** native to the River Amur in Siberia where it occurs in slow flowing water and oxbow lakes. Introduced widely to other parts of the world for aquaculture; it now occurs in some lakes and reservoirs in south east Europe. It has also been introduced to several parts of the United States and appears to be established in the Missouri River. **Reproduction** June–August, but most populations in Europe are maintained by the release of stock raised on fish farms; spawning occurs at 25–30°C and the semipelagic eggs take 30–40 hours to hatch; maturity is reached after 6–7 years (Rumania). **Food** mainly plankton, especially zooplankton but some phytoplankton which it filters by means of its gill rakers which are coarser than those of the Silver Carp. **Value** of considerable commercial value in parts of Europe and other areas where it has been introduced. **Conservation Status** Alien.

Iberocypris: only one species of this genus occurs in Europe:

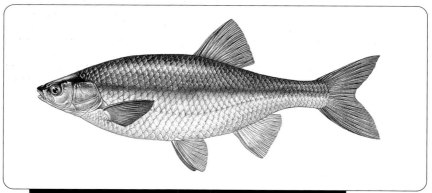

Iberian Carpdace *Iberocypris palaciosi*

Size 15–18 cm; maximum 20 cm. **Distinctive features** small head with inferior mouth; scales small, 45–53 along lateral line; dark iridescent band from head to tail. **Distribution** found only in Spain in a restricted part of the Guadaquivir River system. **Reproduction** uncertain, but probably early summer. **Food** little known of the biology of this species. **Value** no commercial or sporting value. **Conservation Status** Endangered.

Ladigesocypris: only one species of this genus occurs in Europe:

Rhodos Carplette *Ladigesocypris ghigii*

Size 6–9 cm; maximum 10 cm. **Distinctive features** large eyes; lateral line ends below dorsal fin; 25–35 lateral scales. **Distribution** found only in Greece and Turkey in marshes, springs and slow flowing rivers. It has disappeared from a number of systems in which it used to occur. **Reproduction** little known, but believed to be a multiple spawner which may live for up to 5 years. **Food** invertebrates, especially zooplankton and insect larvae; some plant material. **Value** of no commercial or sporting value. **Conservation Status** Vulnerable.

Leucaspius: only one species of this genus occurs in Europe:

Belica *Leucaspius delineatus*

Size 5–9 cm; maximum 12 cm. **Distinctive features** mouth terminal; lateral line incomplete, extending over first 2–13 scales only; 40–46 lateral scales; dorsal fin with 8 and ventral fin with 10–13 branched rays; keel between pelvic and anal fins. **Distribution** found in well vegetated small ponds, and the lower reaches of some rivers in central and eastern Europe from France in the west to the Volga basin in the east; and from Sweden to Austria. Recently introduced to and established in several parts of England. **Reproduction** April–June, spawning in shoals among weeds in shallow water, where the females (which have a tubular genital fold) lay spiral strings of eggs around plants, which are subsequently guarded by males. Matures after 2 years and lives only about 5 years. **Food** invertebrates (especially crustaceans and some insect larvae) and fine plant material. **Value** of local commercial importance in net fisheries in parts of Russia. Rarely angled for but occasionally kept in aquaria. **Conservation Status** Lower Risk.

Carp Family Cyprinidae

Leuciscus: this is a very large genus with considerable uncertainty remaining over the status of the various species and subspecies. 21 species are recognised in the species list for Europe but only 14 of these are discussed fully below. Considerable work remains to be done on the taxonomy of those which are not dealt with in full here (i.e. *Leuciscus burdigalensis, Leuciscus carolitertii, Leuciscus keadicus, Leuciscus lucumonis, Leuciscus montenigrinus, Leuciscus muticellus, Leuciscus zrmanjae*); those covered in full may be identified as follows:

1	Dark band along side of body	*Go to* **2**
	No dark band along side of body	*Go to* **5**
2	Lateral band violet	***Leuciscus souffia*** p126
	Lateral band dull grey or brown	*Go to* **3**
3	Lateral band a thin dark line from eye to base of tail	***Leuciscus polylepis*** p125
	Lateral band broad	*Go to* **4**
4	Lateral band rather faint; 62–64 scales on lateral line	***Leuciscus ukliva*** p128
	Lateral band broad and dark; 70–72 scales on lateral line	***Leuciscus turskyi*** p127
5	More than 54 scales on lateral line	*Go to* **6**
	Less than 55 scales on lateral line	*Go to* **7**
6	55–61 scales on lateral line; 12–14 rays in dorsal fin	***Leuciscus idus*** p123
	73–75 scales on lateral line; 11 rays in dorsal fin	***Leuciscus microlepis*** p125
7	More than 46 scales on lateral line	*Go to* **8**
	Less than 47 scales on lateral line	*Go to* **10**
8	Dorsal fin with 10–11 rays; 10–12 rays in anal fin	***Leuciscus leuciscus*** p124
	Dorsal fin with 12–14 rays; 12–13 rays in anal fin	*Go to* **9**
9	Lateral line with 48–49 scales; 12–14 rays in dorsal fin; scales with dark spots	***Leuciscus svallize*** p127
	Lateral line with 49–54 scales; 12 rays in dorsal fin; scales with dark edges	***Leuciscus illyricus*** p124
10	Anal fin truncated or only slightly emarginate; 43–45 scales on lateral line	***Leuciscus danilewskii*** p122
	Anal fin rounded at tip; if, exceptionally, truncated, then 37–40 scales on lateral line	*Go to* **11**
11	43–46 scales on lateral line; adults large, usually more than 15 cm long	***Leuciscus cephalus*** p122
	36–43 scales on lateral line; adults small, usually less than 15 cm long	*Go to* **12**
12	Lateral line with 36–40 scales; length of head usually less than depth of body	***Leuciscus borysthenicus*** p121
	Lateral line with 40–43 scales; length of head usually greater than depth of body	*Go to* **13**
13	Free edge of anal fin slightly rounded	***Leuciscus aphipsi*** p121
	Free edge of anal fin straight	***Leuciscus pyrenaicus*** p126

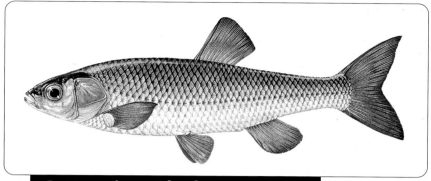

Caucasian Chub *Leuciscus aphipsi*

Size 12–15 cm; maximum 18 cm. **Distinctive features** body with large silvery scales, 40–43 along lateral line; dorsal fin with 8 and anal fin, which is slightly rounded, with 9 branched rays. **Distribution** found only in the Rivers Afips and Psekups (mountainous tributaries of the River Kuban in western Caucasia). **Reproduction** May, spawning over stones and gravel in running water. Males bear tubercles during spawning period. Maturity is usually reached at 3–4 years. **Food** invertebrates, especially insect larvae. **Value** of no commercial or sporting value. **Conservation Status** Vulnerable.

Black Sea Chub *Leuciscus borysthenicus*

Size 15–35 cm; maximum 40 cm. **Distinctive features** mouth terminal; body with large silvery scales, 36–40 along lateral line; dorsal fin with 8–9 and anal fin, which is squarish, with 9–10 branched rays. **Distribution** found in small lakes and in the slow-flowing lower reaches of rivers (e.g. Dniester, Dnieper) entering the Black Sea. **Reproduction** May–June, spawning among weed and stones in slow-moving water. Adults mature after 3–4 years. On average about 2,500 eggs per female (this number is for small fish in the River Kuban) though fecundity will be greater in larger females. **Food** invertebrates when young, invertebrates and small fish when older. **Value** of little commercial or sporting value. **Conservation Status** Lower Risk.

Burdig Dace *Leuciscus burdigalensis*
recorded from France (type locality: River Gironde). **Conservation Status** Vulnerable.

Carol Dace *Leuciscus carolitertii*
recorded from Spain (type locality: River Cega). **Conservation Status** Vulnerable.

Common Chub *Leuciscus cephalus*

Size 30–50 cm; maximum 80 cm; British rod record 3.912 kg (1994). **Distinctive features** snout blunt; forehead wide and flat; body with large silvery scales, less than 46 along lateral line; dorsal fin with 7–9 and anal fin, which is rounded, with 8–9 branched rays. **Distribution** found in running water (occasionally in lakes) throughout most of central and southern Europe. Occurs sometimes in brackish water (e.g. Baltic Sea). **Reproduction** April–June laying adhesive eggs among stones and plants in slow-flowing water. The eggs hatch in 6–8 days and the adults are mature after 3–4 years. 50,000–200,000 eggs per female. **Food** mainly invertebrates and some plant material when young, large invertebrates and fish (occasionally frogs and young voles) when adult. **Value** only locally of importance as a commercial species, when it is caught by nets. It is of considerable sporting value in many parts of Europe. **Conservation Status** Lower Risk.

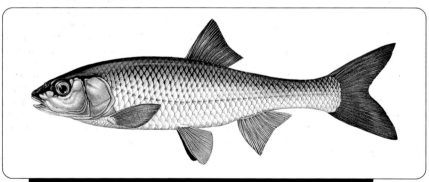

Danilewskii's Dace *Leuciscus danilewskii*

Size 15–20 cm; maximum 25 cm. **Distinctive features** body covered by large silvery scales, 43–45 along the lateral line; dorsal fin with 7 and anal fin with 8–9 branched rays. **Distribution** found only in running water in the basin of the River Don. **Reproduction** March–May, spawning among gravel, stones and weed in streams. Matures after 3–4 years. **Food** invertebrates, especially insects. **Value** of little commercial or sporting value. **Conservation Status** Vulnerable.

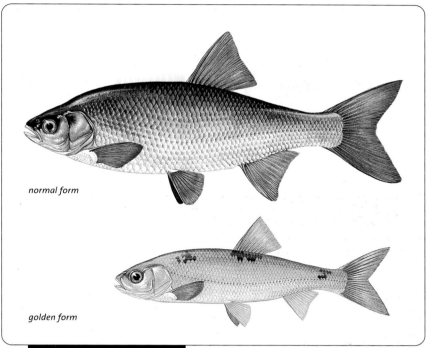

normal form

golden form

Orfe *Leuciscus idus*

Size 35–50 cm; maximum 100 cm; maximum weight up to 8 kg; British rod record 3.033 kg (1994). **Distinctive features** iris yellow; body with well-developed scales, 55–61 along lateral line; dorsal fin with 7–9 and anal fin with 9–10 branched rays. **Distribution** native to most of Europe and parts of Asia east of the Rhine. Introduced successfully to several countries west of the Rhine and also to the United States, where its status is uncertain. Found sometimes in brackish water, it occurs mainly in rivers and some lakes. **Reproduction** April–May, in shallow water among weed and stones. The adhesive eggs hatch in 15–20 days and the adults mature at 5–7 years. 39,000–114,000 eggs per female. **Food** invertebrates, especially molluscs, crustaceans and insect larvae, and some plant material, when young; invertebrates and some fish when larger. **Value** of considerable commercial importance in Russia, where large numbers are taken in nets and traps. Angled for in some parts of Europe. The golden variety is very popular as an aquarium or pond fish. **Conservation Status** Lower Risk.

Yugoslavian Dace *Leuciscus illyricus*

Size 15–20 cm; maximum 25 cm. **Distinctive features** body with large silvery scales, 49–54 along lateral line; dorsal fin with 8 and anal fin with 8–9 branched rays. **Distribution** occurs only in Albania and Croatia where it is found in a variety of waters. **Reproduction** reproductive habits unknown. **Food** invertebrates especially worms, crustaceans and insect larvae. **Value** of minor local commercial (not fishery) and angling importance. **Conservation Status** Vulnerable.

Sparta Dace *Leuciscus keadicus*
recorded from Greece (type locality: River Evrotas). **Conservation Status** Lower Risk.

Common Dace *Leuciscus leuciscus*

Size 15–25 cm; maximum 30 cm; British rod record 574 g (1960). **Distinctive features** inferior mouth; yellowish iris; body with large silvery scales, 49–52 along lateral line; anal fin concave; dorsal fin with 10–11 and anal fin with 8–9 branched rays. **Distribution** found in rivers and streams, occasionally in lakes, throughout Europe except the extreme north-west and south-west. Occasionally in brackish water near river mouths. **Reproduction** March–May, spawning communally to lay pale orange eggs among stones and plants in running water. The adults mature after 3–4 years. 2,500–27,500 eggs per female. **Food** mainly invertebrates, especially aquatic insects – both larvae and adults, including those fallen on to the surface. **Value** of some commercial importance in parts of Russia. A sport species in some parts of western Europe. **Conservation Status** Lower Risk.

Adriatic Dace *Leuciscus svallize*

Size 15–25 cm; maximum 30 cm. **Distinctive features** body with large silvery scales, 48–49 along lateral line; dorsal fin with 8–10 and anal fin with 9–10 branched rays. **Distribution** shoaling fish, found only in clear streams and rivers in Albania, Croatia and Bosnia. **Reproduction** spawns in early spring over stones in running water. **Food** invertebrates, including worms, crustaceans and aquatic flying insects. **Value** of some local commercial importance in net fisheries; also angled for in a few areas. **Conservation Status** Vulnerable.

Turskyi Dace *Leuciscus turskyi*

Size 15–20 cm; maximum 25 cm. **Distinctive features** body with well-developed scales, 70–72 along lateral line; a broad dark lateral band on each side. **Distribution** a shoaling fish, found only in streams and lakes in the Narenta basin in Croatia. **Reproduction** reproductive habits not known. **Food** invertebrates, especially worms, crustaceans and insect larvae. **Value** of little commercial or sporting value. **Conservation Status** Data Deficient.

Ukliva Dace *Leuciscus ukliva*

Size 15–20 cm; maximum 25 cm. **Distinctive features** body with well-developed large scales, 62–64 scales along lateral line; a broad, although rather faint dark line along sides. **Distribution** found only in running water in the Cetina basin in Croatia. **Reproduction** spawns communally over weed and stones. **Food** invertebrates, especially worms, crustaceans and insect larvae. **Value** of little commercial or sporting value. **Conservation Status** Endangered.

Zrmanjan Dace *Leuciscus zrmanjae*
recorded from Croatia (type locality: River Zrmanja). **Conservation Status** Vulnerable.

Pachychilon: Two species of this genus occur in Europe and they can be separated as follows:

1	Absent from the catchments of Lakes Ohrid and Skadar	*Pachychilon macedonicum* p129
	Found native only in the catchments of Lakes Ohrid and Skadar	*Pachychilon pictum* p129

Macedonian Moranec *Pachychilon macedonicum*

recorded from Greece and Macedonia (type locality: Lake Dojran). **Conservation Status** Lower Risk.

Moranec *Pachychilon pictum*

Size 12–15 cm; maximum 16 cm. **Distinctive features** a small slender fish with scattered dark spots; lips very thick and divided into three by deep folds; 43–44 lateral scales; dorsal fin with 11 and anal fin with 11–12 branched rays. **Distribution** found only in two areas of Albania (Ohrid and Skadar basins), but has been introduced and is established in Italy (River Serchio). **Reproduction** shoaling in shallow water during May–August. **Food** invertebrates, especially crustaceans and insect larvae. **Value** of no commercial or sporting value. **Conservation Status** Lower Risk.

Pelecus: only one species of this genus occurs in Europe:

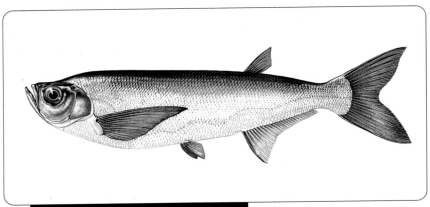

Chekhon *Pelecus cultratus*

Size 30–50 cm; maximum 60 cm; maximum weight 3.5 kg. **Distinctive features** body elongate but strongly compressed laterally; back straight; lateral line undulating along sides; scales small, 90–115 along lateral line; well-developed keel along abdomen, from throat to anus; dorsal fin with 7–8 and anal fin with 24–28 branched rays. **Distribution** found in central and eastern Europe in the basins of the Baltic and northern Black, Caspian and Aral Seas, occurring mainly in the lower reaches and estuaries of large rivers but also in a few lowland lakes. **Reproduction** May–July, communal spawning in both fresh and brackish water. The eggs hatch in 3–4 days and the young mature in 3–4 years, living up to 11 years. 10,000–58,000 eggs per female. **Food** when young, invertebrates – especially crustaceans, insect larvae and adults. When adult, larger invertebrates and small fish (e.g. herring and gobies). **Value** of considerable commercial importance in parts of south-eastern Europe; caught largely in nets and traps. As with the Bleak, its scales are used in some places to coat artificial pearls. **Conservation Status** Lower Risk.

Phoxinellus: this is an important genus with considerable uncertainty remaining over the status of the various species and subspecies. 10 species are recognised in the species list for Europe but only 6 of these are discussed fully below. Considerable work remains to be done on the taxonomy of those which are not dealt with in full here (i.e. *Phoxinellus fontinalis, Phoxinellus metohiensis, Phoxinellus pleurobipunctatus, Phoxinellus prespensis*); those covered in full may be identified as follows:

1	Most of body without scales	Go to **2**
	Most of body covered in thin or rudimentary scales	Go to **3**
2	No scales on body except along lateral line	***Phoxinellus alepidotus*** p132
	Some scales on body although not on back, or caudal peduncle	***Phoxinellus ghetaldii*** p133
3	Body covered by thin fragile scales	***Phoxinellus epiroticus*** p133
	Body covered with rudimentary scales	Go to **4**
4	No dark spots along sides	***Phoxinellus croaticus*** p132
	Dark spots along sides	Go to **5**
5	Relatively few brown spots, rarely below lateral line	***Phoxinellus pstrossi*** p134
	Relatively many dark brown spots, frequently below lateral line	***Phoxinellus adspersus*** p131

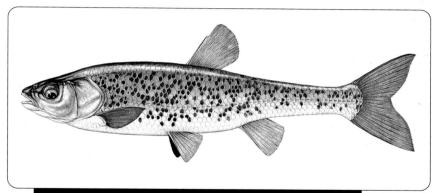

Spotted Minnow *Phoxinellus adspersus*

Size 8–9 cm; maximum 10 cm. **Distinctive features** dark spots along sides, many of them below lateral line. **Distribution** found only in some waters in Croatia. **Reproduction** spawning behaviour unknown. **Food** invertebrates, especially worms, crustaceans and insect larvae. **Value** of no commercial or sporting value. **Conservation Status** Data Deficient.

Adriatic Minnow *Phoxinellus alepidotus*

Size 8–9 cm; maximum 10 cm. **Distinctive features** mouth terminal; body virtually without scales except for a few along the lateral line which is variable in extent. **Distribution** found only in a few basins in Croatia and Bosnia draining into the eastern Adriatic Sea. **Reproduction** spawns over stones in running water. **Food** invertebrates, especially small crustaceans and insect larvae. **Value** of no commercial or sporting significance. **Conservation Status** Vulnerable.

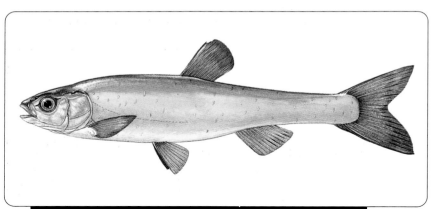

Croatian Minnow *Phoxinellus croaticus*

Size 8–9 cm; maximum 10 cm. **Distinctive features** body with rudimentary scales, but no dark spots along sides. **Distribution** found only in springs and streams in south Croatia, in the basin of the River Zrmanja. **Reproduction** spawns over stones in running water. **Food** invertebrates (mainly crustaceans and insect larvae) and some detritus. **Value** of no commercial or sporting value. **Conservation Status** Vulnerable.

Greek Minnow *Phoxinellus epiroticus*

Size 9–10 cm; maximum 12 cm. **Distinctive features** body covered with thin fragile scales. **Distribution** found only in Lake Janina and the River Luro (Albania, Macedonia and Greece). **Reproduction** spawns over stones in running water. **Food** invertebrates, including crustaceans and insect larvae. **Value** of no sporting or commercial value. **Conservation Status** Vulnerable.

Cave Minnow *Phoxinellus fontinalis*
recorded from Croatia (type locality: Hrnjakova Cave). **Conservation Status** Endangered.

Dalmatian Minnow *Phoxinellus ghetaldii*

Size 9–12 cm; maximum 13 cm. **Distinctive features** most of body without scales, but some present on body, although not on back or caudal peduncle. **Distribution** found only in springs, streams and cave waters in the Popovo basin (Croatia and Bosnia). **Reproduction** spawns over stones in running water. **Food** invertebrates, mainly crustaceans. **Value** of no commercial or sporting value. **Conservation Status** Vulnerable.

Carp Family Cyprinidae

Karst Minnow *Phoxinellus metohiensis*
recorded from Bosnia and Croatia (type locality: various karst waters). **Conservation Status** Data Deficient.

Speckled Minnow *Phoxinellus pleurobipunctatus*
recorded from Greece (type locality: Trichonis Lake and various streams). **Conservation Status** Lower Risk.

Prespa Minnow *Phoxinellus prespensis*
recorded from Greece and Macedonia (type locality: Prespa Lakes). **Conservation Status** Vulnerable.

South Dalmatian Minnow
Phoxinellus pstrossi

Size 8–9 cm; maximum 10 cm. **Distinctive features** a few dark spots along sides, but mostly above lateral line. **Distribution** found only in certain coastal areas of south-west Yugoslavia. **Reproduction** spawning behaviour unknown. **Food** invertebrates, including crustaceans and insect larvae. **Value** of no commercial or sporting value. **Conservation Status** Endangered.

Phoxinus: this genus used to contain *Phoxinus perenurus* which is now placed in the genus *Eupallasella*. Two species now occur in Europe and they may be identified as follows:

1	Large vague dark spots on body, sometimes forming a longitudinal pattern or stripe; no distinct dark speckles on sides; horny tubercles present on heads of mature males	***Phoxinus phoxinus*** p135
	Small distinct dark speckles on body, but no large vague dark spots; horny tubercles never present on heads of mature males	***Phoxinus czekanowskii*** p135

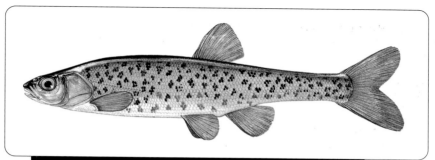

Poznan Minnow *Phoxinus czekanowskii*

Size 5–8 cm; maximum 12 cm. **Distinctive features** scales small, 90–94 along sides; dorsal fin with 7 and anal fin with 6–7 branched rays; body with numerous dark spots; males have no tubercles. **Distribution** found in rivers and lakes in many basins whose waters enter the Arctic Ocean. **Reproduction** June–July, among weed in both running and standing water. **Food** invertebrates, especially crustaceans and insect larvae. **Value** of some commercial value in certain areas of Russia. **Conservation status** lower risk.

♂

♀

Common Minnow *Phoxinus phoxinus*

Size 6–10 cm; maximum 14 cm; British rod record 23 g (1981). **Distinctive features** scales small; 80–100 along lateral line, which may be incomplete near tail; dorsal with 7, anal with 6–7 branched rays; numerous brown and black blotches along sides, sometimes uniting to form stripes; males, with prominent tubercles, brightly coloured (red, black and green) during spawning. **Distribution** found in rivers and lakes over almost the whole of Europe and northern Asia. **Reproduction** May–July, spawning in shoals over stones and gravel (to which the eggs adhere) in running water. The eggs hatch in 5–10 days and the adults mature after 2–3 years, but may live up to 8 years. 200–1,000 eggs per female. **Food** invertebrates, especially crustaceans and insect larvae, and some plant material. **Value** of commercial importance in parts of Russia. A valuable bait species elsewhere. Also used commonly as an aquarium and laboratory fish. **Conservation Status** Lower Risk.

Pimephales: only one member of this introduced genus is established in Europe:

Fathead Minnow *Pimephales promelas*

Size 5–9 cm; maximum 10 cm. **Distinctive features** Body stout; mouth terminal; lateral line incomplete; 42–48 lateral scales; breeding males with fleshy pad behind head. **Distribution** Native to North America, occurring there in still rich lakes or slow flowing streams. Now widely sold in Europe as an aquarium and bait fish. Introduced and now established in Belgium, France and Germany. **Reproduction** The male cleans the underside of a lily leaf or stone with its head pad and spawns there with one or more females during June to August, guarding and cleaning the eggs until they hatch in 5 days at 25°C. Young mature rapidly and may spawn the same year at 5–7 cm. Lives only 2–3 years. **Food** invertebrates, mainly zooplankton and insect larvae, but also algae and vegetable detritus. **Value** of commercial value for aquaria, as bait and as a laboratory test animal. The golden variety is very popular as an aquarium or pond fish. **Conservation Status** Alien.

Pseudophoxinus: three species of this genus are found in Europe. Two of them are considered in detail below (the third species is Pseudophoxinus beoticus) and may be separated as follows: •

---

1   Lateral stripe dark, running full length of body;
    4–6 scales along lateral line

    Pseudophoxinus minutus p137

    Lateral stripe pale bluish; 2–13 scales along lateral line

    Pseudophoxinus stymphalicus p138

Yliki Minnowcarp *Pseudophoxinus beoticus*
recorded only from Greece (type locality: Lake Yliki). **Conservation Status** Vulnerable.

Albanian Minnowcarp
Pseudophoxinus minutus

Size 4–5 cm; maximum 6 cm. **Distinctive features** lateral line very short (extending to only 4–6 scales); ventral fin with 8 rays; dark stripe along each side. **Distribution** found only in Lake Ohrid (Albania and Macedonia). **Reproduction** spawning habits, etc. unknown. **Food** mainly zooplankton, especially crustaceans. **Value** of no commercial or sporting significance. **Conservation Status** Data Deficient.

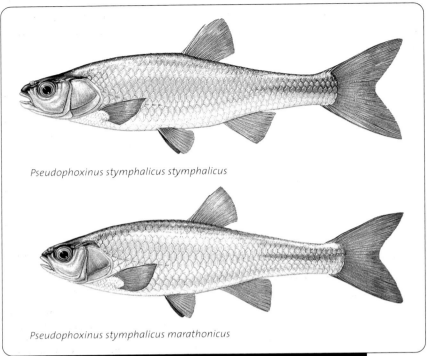

Pseudophoxinus stymphalicus stymphalicus

Pseudophoxinus stymphalicus marathonicus

Marida Minnowcarp
Pseudophoxinus stymphalicus

Size 8–10 cm; maximum 12 cm. **Distinctive features** lateral line incomplete, extending only to first 2–13 scales; anal fin with less than 12 rays. *Pseudophoxinus marathonicus* is now considered to be a subspecies of this species. **Distribution** found only in springs and streams in Morea in southern Greece. **Reproduction** December–March, spawning over aquatic plants. **Food** invertebrates, mainly crustaceans, insect larvae and aerial insects; some plant material. **Value** of little commercial or sporting value. **Conservation Status** Lower Risk.

Pseudorasbora: only one species of this introduced genus is established in Europe:

False Harlequin *Pseudorasbora parva*

Size 4–6 cm; maximum 7 cm. **Distinctive features** mouth slightly superior and oblique; lateral line straight; 34–38 lateral scales; narrow dark line from head to tail. **Distribution** introduced from Asia, where it is native to the Amur River. Now established in many parts of central Europe where it occurs in slow flowing rivers and lakes. **Reproduction** spawns during summer at 21–26°C. **Food** invertebrates, especially planktonic crustaceans, and fish fry. **Value** of little commercial or angling importance, though it has been intentionally stocked in some waters as a forage food for piscivorous fish. Sometimes kept by aquarists. **Conservation Status** Alien.

Rhodeus: only one species of this genus occurs in Europe:

♂

♀

Bitterling *Rhodeus sericeus*

Size 5–8 cm; maximum 10 cm. **Distinctive features** body deep and compressed laterally; scales large, 32–40 laterally; lateral line incomplete; dorsal fin with 9–10 and anal fin with 8–10 branched rays; female bears long ovipositor; male colourful at spawning time. **Distribution** found in rich, slow-flowing rivers and lakes throughout much of the middle of Europe from France eastward to the Caspian Sea. Introduced successfully to a number of countries, such as England. It has also been introduced to the United States where its status is uncertain. **Reproduction** April–June, the fish forming pairs and depositing large yellow eggs into the mantle cavity of large freshwater mussels, by means of an elongate ovipositor. The eggs hatch in 15–20 days and the young leave the mussel in a few days. The young mature after 2–3 years and may live up to 5 or more years. 40–100 eggs per female. **Food** both plants (mainly filamentous and other attached algae), and invertebrates (especially insect larvae and small crustaceans) are eaten. **Value** of little commercial value, but often used as a bait fish and very commonly kept in aquaria. **Conservation Status** Vulnerable.

Rutilus: this is a very large genus with considerable uncertainty remaining over the status of the various species and subspecies. 17 species are recognised in the species list for Europe but only 8 of these are discussed fully below. Considerable work remains to be done on the taxonomy of those which are not dealt with in full here (i.e. *Rutilus arcasii, Rutilus aula, Rutilus basak, Rutilus heckelii, Rutilus karamani, Rutilus meidingeri, Rutilus ohridanus, Rutilus prespensis, Rutilus ylikiensis*); those covered in full may be identified as follows:

1	Dorsal fin with 11 or more rays; normally with a coloured band along side of body	*Go to* **2**
	Dorsal fin with 10 or less rays; rarely with a coloured band along side	*Go to* **5**
2	Lateral line with 55 or more scales	***Rutilus frisii*** p142
	Lateral line with 50 or less scales	*Go to* **3**
3	Pectoral fins with 16 rays	***Rutilus rutilus*** p147
	Pectoral fins with 17–18 rays	*Go to* **4**
4	Pharyngeal teeth 6.5; no marks along sides of body	***Rutilus pigus*** p145
	Pharyngeal teeth 5.5; often with greyish stripes along sides of body	***Rutilus rubilio*** p146
5	Pharyngeal teeth usually 4.4	***Rutilus macedonicus*** p144
	Pharyngeal teeth usually 5.5	*Go to* **6**
6	Light grey band along side of body	***Rutilus macrolepidotus*** p144
	Marked dark band along side of body	*Go to* **7**
7	Dorsal fin concave	***Rutilus lemmingii*** p143
	Dorsal fin convex	***Rutilus alburnoides*** p141

Calandino Roach *Rutilus alburnoides*

Size 12–15 cm; maximum 20 cm. **Distinctive features** marked dark band along each side; dorsal fin convex. **Distribution** found only in rich slow-flowing or standing waters in Portugal and south-west Spain. **Reproduction** April–May, spawning communally among weed and stones in shallow water, where the adhesive yellow eggs are laid. **Food** invertebrates, notably worms and insect larvae. **Value** of little commercial or sporting significance. **Conservation Status** Lower Risk.

Carp Family Cyprinidae

Galician Roach *Rutilus arcasii*
recorded from Portugal and Spain (type locality: River Cailes, etc.). **Conservation Status** Lower Risk.

Alpine Roach *Rutilus aula*
recorded from Italy (type locality: Venetian Province) and Switzerland. **Conservation Status** Lower Risk.

Dalmatian Roach *Rutilus basak*
recorded from Croatia (type locality: Lake Drusino). **Conservation Status** Vulnerable.

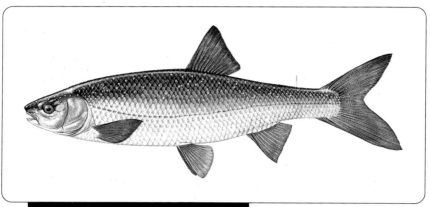

Pearl Roach *Rutilus frisii*

Size 40–50 cm; maximum 71 cm; maximum weight about 7.1 kg. **Distinctive features** body with large silvery scales, 60–67 scales along lateral line. **Distribution** found in a few areas of Germany and Austria in the Upper Danube Basin, in basins entering the Black Sea, the Sea of Azov and the Caspian Sea. Occurring particularly in the lower reaches of rivers. **Reproduction** April–May, in shallow water among weed or over gravel and stones. The young hatch in about 10 days and mature after 3–5 years. Large migrations occur along rivers, and into and out of lakes at various times of the year. Average about 138,400 eggs per female. **Food** a wide variety of invertebrates, together with some plant material and small fish. **Value** of considerable commercial value in many areas near the Black and Caspian Seas, and many thousands of large specimens are netted annually from some rivers. Also important as a sport species in some regions. **Conservation Status** Lower Risk.

Heckel's Roach *Rutilus heckelii*
recorded from the Black Sea (type locality: Black Sea). **Conservation status** Lower Risk.

Karaman's Roach *Rutilus karamani*
recorded from Albania and Montenegro (type locality: Lake Ohrid). **Conservation status** Vulnerable.

Pardilla Roach *Rutilus lemmingii*

Size 12–15 cm; maximum 20 cm. **Distinctive features** dark band along each side; mouth slightly inferior; lower lip with horny edge; 58–69 lateral scales. Some experts place this species in the genus *Chondrostoma*. **Distribution** in standing and slow-flowing waters in Portugal and south-east Spain. **Reproduction** April–May, spawning in large shoals among weed in shallow water where the adhesive yellow eggs are laid. **Food** invertebrates, especially worms, crustaceans and insect larvae. **Value** of little commercial or sporting importance. **Conservation status** Vulnerable.

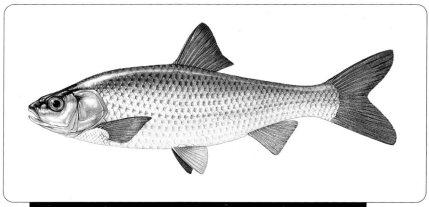

Macedonian Roach *Rutilus macedonicus*

Size 12–15 cm; maximum 18 cm. **Distinctive features** large silvery scales on body and only rarely with a dark band along side; pharyngeal teeth usually 4.4. **Distribution** occurring only in rich slow-flowing or standing waters in a few small basins running into the north-west of the Aegean Sea. **Reproduction** April–May, shoaling among weed in shallow water. **Food** invertebrates, mainly worms, crustaceans and insect larvae, and some plant material. **Value** of no commercial or sporting significance. **Conservation Status** Vulnerable.

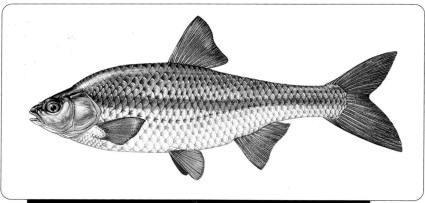

Portuguese Roach *Rutilus macrolepidotus*

Size 12–15 cm; maximum 18 cm. **Distinctive features** large silvery scales on body and a light grey band along each side. **Distribution** found only in a few areas in Portugal, in lakes and lowland rivers. **Reproduction** April–May, shoaling mainly among weed in shallow water where the adhesive yellow eggs are laid. **Food** invertebrates, especially worms, crustaceans and insect larvae, and some plant material. **Value** of little commercial or sporting significance. **Conservation Status** Endangered.

Austrian Roach *Rutilus meidingeri*
recorded from Austria (type locality: Lake Attersee). **Conservation Status** Endangered.

Ohrid Roach *Rutilus ohridanus*
recorded from Macedonia (type locality: Lake Ohrid). **Conservation Status** Vulnerable.

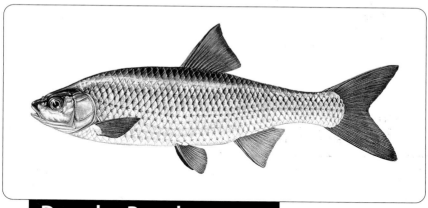

Danube Roach *Rutilus pigus*

Size 20–30 cm; maximum 40 cm; maximum weight about 1 kg. **Distinctive features** rather slender fish with large silvery scales (44–49 along lateral line); dorsal fin with 10–11 and anal fin with 10–13 branched rays. **Distribution** found in slow-flowing or still waters in parts of the Danube basin (subspecies *Rutilus pigus virgo*) and northern Italy (Po basin: subspecies *Rutilus pigus pigus*). **Reproduction** April–May, spawning in shoals among weed, etc. The adults mature after 2–3 years. 35,000–60,000 eggs per female. **Food** invertebrates, especially worms, molluscs and crustaceans, and some aquatic plants. **Value** of little commercial value, it is sometimes sought after as a sport species. **Conservation Status** Endangered.

Prespa Roach *Rutilus prespensis*
recorded from Greece and Macedonia (type locality: Prespa Lakes). **Conservation Status** Vulnerable.

145

Adriatic Roach *Rutilus rubilio*

Size 20–25 cm; maximum 30 cm. **Distinctive features** body with large silvery scales, 36–44 along lateral line, and poorly defined greyish stripes along sides; dorsal fin with 8–9 and anal fin with 8–10 branched rays. **Distribution** found over most of Italy and basins in Slovenia, Croatia and Albania draining into the Adriatic Sea. Occurring mainly in slow-flowing and standing waters. **Reproduction** April–May, spawning in shoals, laying yellowish adhesive eggs among weed in shallow water. **Food** mainly invertebrates, particularly insect larvae and worms. In a few warm waters the main food is filamentous algae. **Value** of some local importance to net fishermen; is also angled for in some places. **Conservation Status** Lower Risk.

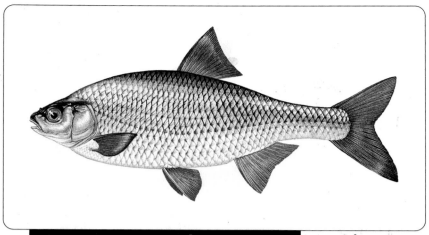

Common Roach *Rutilus rutilus*

Size 20–35 cm; maximum 52 cm; a fish of 44 cm weighed 2.1 kg; British rod record 1.899 kg (1990). **Distinctive features** body with large silvery scales, 42–45 along lateral line; dorsal fin (9–11 branched rays) directly above base of pelvic fins; anal fin with 9–11, pectoral fins with 16 branched rays. **Distribution** common in lakes, canals and slow-flowing rivers over much of northern Europe and Asia, except extreme north. Sometimes found in brackish water near the mouths of large rivers, e.g. in the Baltic and Black Seas. It is common also in parts of the Caspian Sea. **Reproduction** May–June, among weed in shallow water, usually in standing water but sometimes moving into fast-flowing water to spawn. The adhesive yellowish eggs hatch in 5–10 days and the larvae move round in large shoals, as do the adults which mature in about 2–3 years. 1,000–14,600 eggs per female. Hybridises readily with several other cyprinid species. **Food** an omnivorous shoaling species feeding on both aquatic plants (especially attached algae) and invertebrates of many kinds. **Value** commercially important in some parts of eastern Europe where it is caught in traps and nets. An important sport species in many countries where it is caught on set lines, using a variety of baits. Small roach are sometimes used as bait in pike fishing. **Conservation Status** Lower Risk.

Yliki Roach *Rutilus ylikiensis*
recorded from Greece (type locality: Lake Yliki). **Conservation Status** Lower Risk.

Scardinius: this is a small genus but with uncertainty remaining over the status of some species and subspecies. 5 species are recognised in the species list for Europe but only 2 of these are discussed fully below. Considerable work remains to be done on the taxonomy of those which are not dealt with in full here (i.e. *Scardinius acarnanicus, Scardinius racovitzai, Scardinius scardafa*); those covered in full may be identified as follows:

1 Body deep, greatest depth (which is at the dorsal fin) usually about twice depth of head or more; profile not, or hardly, concave just behind head	*Scardinius erythrophthalmus* p148
Body less deep, greatest depth (which occurs well in front of dorsal fin) less than twice depth of head; profile clearly concave behind head	*Scardinius graecus* p149

Acheloos Rudd *Scardinius acarnanicus*
recorded Greece (type locality: Lake Trichonis, etc.). **Conservation Status** Lower Risk.

Common Rudd *Scardinius erythrophthalmus*

Size 15–30 cm; maximum 45 cm; British rod record 2.041 kg (1933). **Distinctive features** iris gold; body with large silvery scales, 40–45 along lateral line; keel between pectoral fins and anus; lower fins bright red; origin of pelvic fins anterior to that of dorsal fin; dorsal fin with 8–9 and anal fin with 10–11 branched rays. **Distribution** found in lakes and slow-flowing rivers throughout much of Europe, occasionally in brackish water. Widely introduced as an ornamental or sport fish it is now established in several countries, including the United States. **Reproduction** April–June, spawning in shoals among weed. The adhesive eggs hatch in 5–8 days and the adults mature after 2–3 years. 90,000–200,000 eggs per female. **Food** invertebrates (especially molluscs and insect larvae) and aquatic vegetation; some feeding on insects trapped on the surface. **Value** of little commercial value, but of value as a sport and a bait species in some countries. The golden variety is commonly kept in aquaria and ponds. **Conservation Status** Lower Risk.

Greek Rudd *Scardinius graecus*

Size 20–35 cm; maximum 40 cm. **Distinctive features** body with large silvery scales, greatest body depth anterior to dorsal fin; keel between pectoral fins and anus. **Distribution** found only in lakes (Yliki and Paralimni) and slow-flowing waters (River Kifissos) in southern Greece. **Reproduction** April–June, spawning in shoals among weed in shallow water. **Food** mainly zooplankton when young, invertebrates and various plant species later. **Value** of some local importance as both a commercial and a sport species. **Conservation Status** Lower Risk.

Rumanian Rudd *Scardinius racovitzai*
recorded from Rumania (type locality: Petzea Pond). **Conservation Status** Vulnerable.

Adriatic Rudd *Scardinius scardafa*
recorded from Albania, Bosnia, Croatia, Italy (type locality: Lake Nemi, etc.) and Montenegro. **Conservation Status** Vulnerable.

Carp Family Cyprinidae

Tinca: only one species of this genus occurs in Europe:

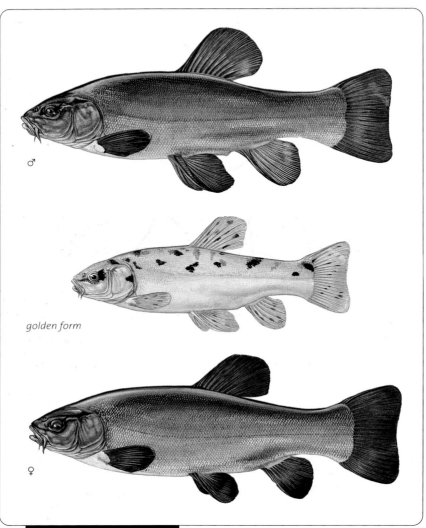

♂

golden form

♀

Tench *Tinca tinca*

Size 20–40 cm; maximum 65 cm; British rod record 6.548 kg (1993). **Distinctive features** one pair of small barbels at mouth, which is terminal; eyes reddish; fins very rounded; body deep, with small, deeply embedded scales, 95–120 along lateral line; dorsal fin with 8 and anal fin with 6–8 branched rays. **Distribution** found in rich, weedy lakes and slow-flowing rivers over most of Europe except northern areas; occasionally in brackish water in eastern Baltic. Widely introduced to other continents, it is now established in several parts of North America. **Reproduction** May–July, laying greenish eggs among weed in shallow water. The eggs hatch in 3–5 days, and adults mature after 3–4 years. They may live up to 10 years. 280,000–827,000 eggs per female. **Food** invertebrates, especially bottom-living molluscs, crustaceans and insect larvae. **Value** of little commercial importance, but valued as a sport species in many countries where it is caught by baited hooks of various types. The golden variety is a popular aquarium and pond fish. **Conservation Status** Lower Risk.

Tropidophoxinellus: there is considerable doubt over the validity of this genus; two species of which occur in Europe.

Hellenic Minnowroach
Tropidophoxinellus hellenicus

Size 5–10 cm. **Distinctive features** less than 10 branched anal rays; premaxilla with long ascending process; small knob on dentary bone. **Distribution** found only in rivers and lakes within the Pinios and Acheloos River systems in Greece. **Reproduction** April–July, spawning over aquatic plants. **Food** invertebrates and some plant material. **Value** of limited local commercial importance. **Conservation Status** Lower Risk.

Spartian Minnowroach
Tropidophoxinellus spartiaticus

Size 5–10 cm. **Distinctive features** less than 10 branched anal rays; premaxilla with long ascending process; small knob on dentary bone. **Distribution** found only in springs and streams in Greece. **Reproduction** not known. **Food** invertebrates and some plant material. **Value** of no commercial or angling value. **Conservation Status** Vulnerable.

Carp Family Cyprinidae

Vimba: three species of this genus occur in Europe. Two species are dealt with in full here, the other being *Vimba elongata*:

1	Marked keel behind dorsal fin	*Go to **2***
	No keel behind dorsal fin	***Vimba melanops*** p152
2	Maximum depth of body more than 25% of length	***Vimba vimba*** p153
	Maximum depth of body less than 25% of length	***Vimba elongata*** p152

River Vimba *Vimba elongata*
recorded from Austria (type locality: River Danube). **Conservation Status** Vulnerable.

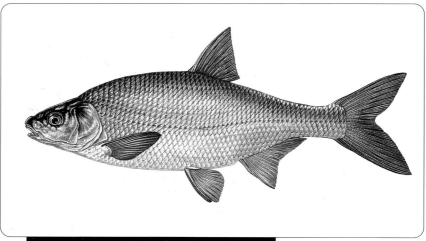

Dark Vimba *Vimba melanops*

Size 20–30 cm; maximum 40 cm. **Distinctive features** inferior mouth; body laterally compressed; head relatively larger than *Vimba vimba*; 52–58 scales along lateral line; 42–43 vertebrae. **Distribution** found in Greece, Bulgaria, Macedonia and Turkey, in lakes and slow flowing rivers. **Reproduction** spawning during summer over areas of stones and gravel. **Food** invertebrates and some plant material. **Value** of some local commercial and angling importance. **Conservation Status** Vulnerable.

Common Vimba *Vimba vimba*

Size 20–30 cm; maximum 50 cm; specimens of average length 30 cm weigh about 490 g. **Distinctive features** sturdy, laterally compressed fish with 53–61 scales along lateral line; a scaleless abdominal keel before vent, and a scaled keel behind dorsal fin; dorsal fin with 8–9 and anal fin with 17–21 branched rays. **Distribution** found in the middle and lower reaches of rivers, in some lakes and in basins linked to the Baltic, Black and Caspian Seas, in whose waters it also occurs. **Reproduction** May–July, migrating upstream and spawning among weed and stones in shallow running water. The eggs hatch in 5–10 days and the young mature after 3–4 years. 27,000–115,500 eggs per female. Hybridises with *Abramis*. **Food** benthic invertebrates, especially worms, molluscs and insect larvae. **Value** of considerable commercial value in eastern Europe and caught in large numbers in nets and traps. **Conservation Status** Lower Risk.

Spined Loach Family
Cobitidae

The Cobitidae, or spined loaches, are all freshwater, bottom-living fish, found only in Europe and Asia. They occur in ponds, lakes and running waters of most kinds. There are many genera and species, particularly in southern tropical Asia. Together with the closely related Balitoridae they include about 150 species.

The body is characteristically very elongate, and usually cylindrical or only slightly compressed laterally. Scales may be absent, but where present they are normally very small and embedded within the skin. The eyes are small and usually rather dorsal in position, whilst the mouth is inferior with fleshy lips, and surrounded by 6 to 12 sensory barbels. The Cobitidae are related to the Cyprinidae, and possess pharyngeal teeth for crushing food. In some genera the head is unusual in that the anterior part articulates with the rest. There is a single dorsal fin, and all the fins are moderately developed.

Many members of the family are able to supplement inadequate oxygen supplies in stagnant water by swimming to the surface and swallowing air. Oxygen is absorbed from this in the gut, especially the hind portion, and the remainder is passed out through the anus. Some species, which can breathe

Ventral views of head and mouth

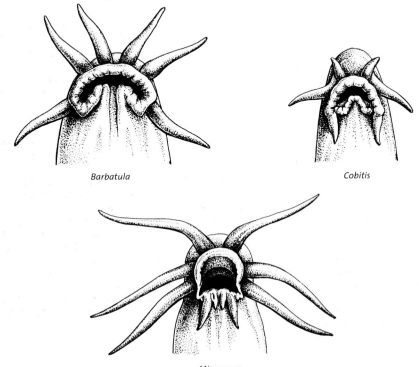

Barbatula

Cobitis

Misgurnus

154

Key to Cobitidae in Europe

1	Either 6 or 8 barbels	*Go to* **2**
	10 barbels (4 on lower jaw)	**Misgurnus fossilis** p160
2	Colour pattern on sides including one or more distinct rows of longitudinal spots	*Go to* **3**
	Colour pattern on sides mottled with no distinct rows of spots	*Go to* **8**
3	Markings mainly large irregular blotches	**Cobitis maroccana** p158
	Markings smaller and in lines	*Go to* **5**
4	No distinct dark line joining spots in lateral row	*Go to* **5**
	Distinct dark line joining spots in lateral row	*Go to* **6**
5	Caudal fin with 14 branched rays; speckles above main lateral row also tending to form a distinct row	**Cobitis taenia** p159
	Caudal fin with less than 14 branched rays; speckles above main lateral row not forming a distinct row	**Sabanejewia balcanica** p160
6	Spotting relatively light; area above lateral row of spots forming a mainly clear band	*Go to* **7**
	Spotting relatively dark; many small spots in area above lateral row	**Cobitis elongata** p157
7	Clear band above lateral spots broad	**Cobitis meridionalis** p158
	Clear band above lateral spots narrow	**Sabanejewia romanica** p161
8	Body coloration mainly in large patches	*Go to* **9**
	Body coloration mainly as small dark speckles	*Go to* **10**
9	Dark triangle formed by two halter-shaped stripes in front of eyes, no obvious elongate spots in front of dorsal fin	**Sabanejewia larvata** p161
	No dark triangle in front of eyes; two elongate spots in front of dorsal fin	**Cobitis conspersa** p157
10	Sides of body mainly with small speckles; suborbital spine strong, its branches similar in length	**Cobitis caucasica** p156
	Obvious dark streak running laterally on body; branches of suborbital spine greatly differing in length	**Cobitis caspia** p156

in this way (and presumably often have quantities of air inside them), are sensitive to changes in barometric pressure and become very active during rapidly decreasing atmospheric pressure preceding a storm.

Most of the species are nocturnal in habit, remaining hidden during daylight under stones, among weed or buried in sand or gravel. The majority feed on benthic invertebrates, but a number are adapted for browsing on algae and various microscopic animals.

This is a small but important family, with uncertainty remaining over the status of some species and subspecies. 24 species are recognised in the species list for Europe but only 11 of these are discussed fully below. Considerable work remains to be done on the taxonomy of those which are not dealt with in full here (i.e. *Cobitis arachthosensis, Cobitis bilineata, Cobitis calderoni, Cobitis elongatoides, Cobitis hellenica, Cobitis megaspila, Cobitis paludica, Cobitis punctilineata, Cobitis stephanidisi, Cobitis strumicae, Cobitis trichonica, Cobitis vardarensis, Sabanejewia bulgarica*); the status of one species introduced from Asia, *Misgurnus anguillicaudatus* (Cantor 1842), not included in the checklist, is uncertain. Those covered in full may be identified as follows:

Spined Loach Family Cobitidae

Arachthos Loach *Cobitis arachthosensis*
recorded recently from Greece. **Conservation Status** Data Deficient.

Twostriped Loach *Cobitis bilineata*
recorded from Italy (type locality: Modenese and Treviso). Possibly a subspecies of *Cobitis taenia*. **Conservation Status** Lower Risk.

Lamprehuela Loach *Cobitis calderoni*
recorded from Portugal and Spain (type locality: Arlanzon Stream). **Conservation Status** Vulnerable.

Caspian Loach *Cobitis caspia*

Size 4–5 cm; maximum 7 cm. **Distinctive features** head compressed laterally; 6 barbels around the mouth; a double spine below each eye, the branches of this spine differing greatly in length; scales small; dorsal fin with 6–7 and anal fin with 5 branched rays. **Distribution** found in both fresh and brackish water in the lower reaches of rivers (e.g. Kura and Ural) whose basins are linked to the west of the Caspian Sea. **Reproduction** April–May, among plants, gravel and sand in slow-flowing or even still water, usually in fresh, but sometimes in brackish habitats. **Food** active at night, feeding on bottom-living invertebrates, mainly worms, small molluscs and insect larvae. **Value** of no commercial or sporting value. **Conservation Status** Lower Risk.

Caucasian Loach *Cobitis caucasica*

Size 6–8 cm; maximum 11 cm. **Distinctive features** head compressed laterally; 6 barbels around mouth; a double spine below each eye – the branches of this spine of approximately equal length; dorsal fin with 6–7 and anal fin with 5 branched rays; sides of body marked with numerous small dark speckles. **Distribution** found only in the upper and middle reaches of certain rivers between the Black and Caspian Seas (e.g. Kuma, Sulak and Terek). **Reproduction** May–July, among stones, gravel and plants in running water. **Food** bottom-dwelling invertebrates, especially worms and insect larvae. **Value** of no commercial or sporting value. **Conservation Status** Lower Risk.

Venetian Loach *Cobitis conspersa*

Size 6–8 cm; maximum 9 cm. **Distinctive features** head compressed laterally; 6 barbels around mouth; a double spine under each eye; 2 elongate dark spots in front of dorsal fin. **Distribution** found only in a few rivers in northern Italy (e.g. Brenta and Gua). **Reproduction** April–June, among plants, sand and gravel in shallow running water. **Food** nocturnal, feeding on bottom-dwelling invertebrates, mainly worms and insect larvae, and some algae. **Value** of no commercial or sporting value. **Conservation Status** Lower Risk.

Balkan Loach *Cobitis elongata*

Size 12–15 cm; maximum 17 cm. **Distinctive features** head and caudal peduncle compressed laterally; 6 barbels around the mouth; a small double spine below each eye; scales minute; large round dark spots along sides with a narrow dark line running through them. **Distribution** found only in streams in Yugoslavia and Rumania (e.g. Donau). **Reproduction** April–June, in shallow running water, spawning there on the bottom among stones and gravel. **Food** bottom-living invertebrates, mainly worms, small molluscs and insect larvae. **Value** of no commercial or sporting value. **Conservation Status** Vulnerable.

Slender Loach *Cobitis elongatoides*
recorded from Romania (type locality: Rivers Neajlov, Jiu and Sii). **Conservation Status** Vulnerable.

Hellenic Loach *Cobitis hellenica*
recorded recently from Greece (Louros and Arachthos River basins). **Conservation Status** Lower Risk.

Spined Loach Family Cobitidae

Moroccan Loach *Cobitis maroccana*

Size 10–12 cm; maximum 15 cm. **Distinctive features** head compressed laterally; 6 barbels around mouth; double spine below each eye; characteristic pattern along body. **Distribution** found, in Europe, only in Portugal and Spain, in both slow and fast flowing rivers. **Reproduction** during summer, among stones and aquatic plants. **Food** not known but probably bottom living invertebrates. **Value** of no commercial or angling importance. **Conservation Status** Lower Risk.

Delta Loach *Cobitis megaspila*
recorded from Rumania (type locality: River Danube delta). **Conservation Status** Vulnerable.

Prespa Loach *Cobitis meridionalis*

Size 8–10 cm; maximum 12 cm. **Distinctive features** 6 barbels around mouth, and a movable double spine below each eye; head compressed laterally; characteristic body pattern. **Distribution** found only in the Prespa Lakes (Greece, Albania and Macedonia) and in associated running waters. **Reproduction** April–May, among stones in running water. **Food** bottom invertebrates. **Value** of no commercial or angling importance. **Conservation Status** Vulnerable.

Marsh Loach *Cobitis paludica*
recorded from Spain (type locality: Fuente del Roble). **Conservation Status** Vulnerable.

Striped Loach *Cobitis punctilineata*
recorded recently from Greece (Aggitis River). **Conservation Status** Lower Risk.

Stephanidis' Loach *Cobitis stephanidisi*
recorded from Greece (type locality: Kefalovrissos spring). **Conservation Status** Vulnerable.

Strumica Loach *Cobitis strumicae*
recorded from Greece (type locality: River Strumica and Monospitovo Swamp). **Conservation Status** Lower Risk.

Spined Loach *Cobitis taenia*

Size 8–10 cm; maximum 14 cm. **Distinctive features** 6 barbels around mouth, and a movable double spine below each eye; head compressed laterally; eyes small and high on head; scales minute; dorsal fin with 6–7 and anal fin with 5–6 branched rays; male pectoral fins with thickened second ray. **Distribution** found where the bottom is soft (where it burrows), in slow-flowing streams and lakes over most of Europe, from England east to Asia, except the extreme north. **Reproduction** April–July, among stones and weed in shallow running water. The larvae become bottom living almost immediately after hatching. **Food** active at dusk, feeding on bottom-dwelling invertebrates, especially worms, small molluscs and insect larvae. **Value** of no commercial or sporting value. **Conservation Status** Lower Risk

Spined Loach Family Cobitidae

Trichonis Loach *Cobitis trichonica*
recorded from Greece. (type locality: Lake Trichonis). **Conservation Status** Lower Risk.

Vardar Loach *Cobitis vardarensis*
recorded from Greece and Macedonia (type locality: River Vardar). **Conservation Status** Lower Risk.

Weather Loach *Misgurnus fossilis*

Size 15–30 cm; maximum 50 cm. **Distinctive features** 10 barbels around mouth, largest pair on upper lip; no spine under eye; a large species with an elongate rounded body; 135–175 minute lateral scales; dorsal fin with 5–7 and anal fin with 5–6 branched rays; several dark lateral stripes. **Distribution** found in rich ponds and small lakes on flood plains in many parts of central and northern Europe, from Belgium to the Caspian Sea. Can breathe air if oxygen is low or when habitat is drying up. **Reproduction** April–June, reddish eggs laid among weeds in shallow water. The larvae possess fine external gills for a few days after hatching. 65,000–170,000 eggs per female. **Food** bottom-dwelling invertebrates, especially worms, molluscs and insect larvae. **Value** of little commercial or sporting value, although sometimes kept as an aquarium or pond fish. **Conservation Status** Endangered.

Golden Loach *Sabanejewia balcanica*

Size 8–12 cm; maximum 14 cm. **Distinctive features** 6 barbels around mouth; a double spine beneath each eye; head compressed laterally; scales minute, 170–200 along lateral line; dorsal fin with 6–7 and anal fin with 5–6 branched rays; caudal fin with less than 14 branched rays. Formerly known as *Cobitis aurata*. **Distribution** found only in the upper and middle reaches of streams in river basins emptying into the Black and Caspian Seas (e.g. Danube, Vordov and Kuban). **Reproduction** May–July, spawning among plants, gravel and stones in running water – especially in the upper reaches of streams and rivers. **Food** active at dusk, feeding on bottom-dwelling invertebrates, especially small worms and insect larvae. **Value** of no commercial or sporting value. **Conservation Status** Lower Risk.

Bulgarian Loach *Sabanejewia bulgarica*
recorded from Bulgaria (type locality: River Danube and tributaries). **Conservation Status** Data
Deficient.

Italian Loach *Sabanejewia larvata*

Size 5–8 cm; maximum 9 cm. **Distinctive features** head
compressed laterally; 6 barbels around mouth; a double spine
below each eye; scales small; a dark triangle in front of eyes; 2
dark spots on caudal peduncle. **Distribution** found only in
running water in the Bergantino region of northern Italy.
Reproduction May–June, spawning among plants, stones,
gravel and sand in shallow running water. **Food** mainly
bottom-living worms and insect larvae. **Value** of some commercial and sporting value;
recently used in aquaculture. **Conservation Status** Vulnerable.

Rumanian Loach *Sabanejewia romanica*

Size 8–10 cm; maximum 12 cm. **Distinctive features** 6 barbels
around mouth; a double spine below each eye; head
compressed laterally; scales small; body relatively lightly
marked, a clear band running above the main lateral row of
spots. **Distribution** found only in the upper reaches of certain
tributaries of the Danube in Rumania. **Reproduction** April–July,
spawning on the bottom in shallow running water among
stones and gravel. **Food** bottom-dwelling invertebrates, especially worms, small molluscs
and insect larvae. **Value** of no commercial or sporting value. **Conservation Status**
Vulnerable.

Stone Loach Family
Balitoridae

This family of loaches is closely related to, and was formerly grouped with, the Cobitidae. There are some 400 species in the family, but most of them occur in tropical and subtropical Asia. Only 4 species occur in Europe. The status and nomenclature of another species (*Orthrias pindus* Economidis 1991) is uncertain and it is not covered here. The 4 European species may be distinguished as follows:

Key to Balitoridae in Europe

1	Caudal fin truncated	***Barbatula barbatula*** p163
	Caudal fin emarginate	*Go to* **2**
2	Minimum body depth less than half caudal peduncle length; no dentiform process on the upper jaw	***Barbatula merga*** p163
	Minimum body depth more than half caudal peduncle length; dentiform process usually present on upper jaw	*Go to* **3**
3	Caudal fin slightly emarginate	***Barbatula angorae*** p162
	Caudal fin markedly emarginate	***Barbatula bureschi*** p163

Angora Loach *Barbatula angorae*

Size 6–8 cm; maximum 9 cm. **Distinctive features** 6 barbels around mouth; no spine under eye; head rounded; scales minute; 79 lateral line pits; dorsal fin with 7–8 and anal fin with 5 branched rays; caudal fin slightly emarginate; body slim, minimum depth more than half length of caudal peduncle. Formerly *Cobitis angorae*. **Distribution** found in rivers, streams and some lakes in certain basins entering the Black and Aegean Seas, (e.g. Kamer and Coruh). **Reproduction** May–July, spawning among stones, gravel and plants in shallow running water. Specimens mature when they are longer than 5 cm. **Food** benthic invertebrates, especially worms and insect larvae. **Value** of no commercial or sporting value. **Conservation Status** Vulnerable.

Stone Loach *Barbatula barbatula*

Size 8–12 cm; maximum 18 cm. **Distinctive features** 6 barbels around mouth; no spine under eye; eyes high on head; head rounded; scales very small; dorsal fin with 7 and anal fin with 5 branched rays; caudal fin truncated. Formerly *Noemacheilus barbatulus*. **Distribution** found, often under stones or weed, in streams, rivers and some lakes throughout Europe except the extreme north and south; also into Baltic brackish water. Introduced to Ireland. **Reproduction** April–June, among, or under, stones and weed in running water. The young mature after 2–3 years and may live up to 8 years. 500,000–800,000 eggs per female. **Food** active at night, feeding on benthic invertebrates, especially insect larvae of various kinds. **Value** occasionally eaten, but really of no commercial use. Sometimes used in sport fishing as a bait species. **Conservation Status** Lower Risk.

Struma Loach *Barbatula bureschi*

Size 6–8 cm; maximum 9 cm. **Distinctive features** 6 barbels around mouth; no spine under eye; head rounded; dorsal fin with 8 and anal fin with 4–5 branched rays; caudal fin more emarginate than in *Barbatula angorae*; body slim, minimum depth more than half length of caudal peduncle; characteristic body pattern. Possibly just a subspecies of *Cobitis angorae*. **Distribution** found in rivers, streams and some lakes in certain basins entering the Aegean Sea, e.g. Rivers Struma and Mesta. **Reproduction** May–July, spawning among stones, gravel and plants in shallow running water. Specimens mature when they are longer than 5 cm. **Food** active at night, feeding on benthic invertebrates, especially worms and insect larvae. **Value** of no commercial or sporting value. **Conservation Status** Lower Risk.

Terek Loach *Barbatula merga*

Size 6–8 cm; maximum 10 cm. **Distinctive features** 6 barbels around mouth; no spine under eye; head rounded; scales small but distinct; dorsal fin with 7 and anal fin with 5 branched rays; caudal fin emarginate; body rather deep, its minimum depth less than half the caudal peduncle length. Formerly *Noemacheilus merga*. **Distribution** found in running waters in basins of various rivers (e.g. Kuban, Kuma, Terek) in south-western Russia between the Black and Caspian Seas. **Reproduction** spawning habits little known. **Food** bottom-dwelling invertebrates, especially worms and insect larvae. **Value** of no commercial or sporting value. **Conservation Status** Lower Risk.

American Catfish Family
Ictaluridae

The Ictaluridae, or North American catfishes, are native to rich lakes, ponds and slow-flowing rivers in North and Central America. There are 5 genera with some 25 species, largely restricted to the New World, although a few species have been introduced elsewhere. All are freshwater species. The body is elongate and scaleless, with a flattened head bearing barbels. An adipose fin is present, and Dorsal and pectoral fins are armed with spines.

Several species have been introduced to Europe, mainly for sale to aquarists; and 3 of these are now established.

Key to Ictaluridae in Europe

1	Caudal fin emarginate, not deeply forked	*Go to* **2**
	Caudal fin deeply forked	***Ictalurus punctatus*** p165
2	Barbs on posterior pectoral spines weak or absent; 17–21 anal rays; 17–19 gill rakers; dorsal fin membranes darkened.	***Ameiurus melas*** p164
	Barbs on posterior pectoral spines large; 21–24 anal rays; 13–15 gill rakers; dorsal fin membranes not darkened	***Ameiurus nebulosus*** p165

Black Bullhead *Ameiurus melas*

Size 20–30 cm; maximum 45 cm; maximum weight about 3 kg. **Distinctive features** broad flat head with 8 barbels; barbs on posterior pectoral spines weak or absent; anal fin with 17–21 rays. **Distribution** native to eastern North America, but introduced to Europe (e.g. Italy) where it has established itself successfully. Occurs in rich slow-flowing rivers and lakes. **Reproduction** June–July, in a nest among logs, plants, etc.. Eggs and young are guarded by both parents. 2,550–3,850 eggs per female. Matures at 2–3 years and lives about 9 years. **Food** both plants and invertebrates, especially crustaceans. Small fish are also eaten. **Value** of little commercial or sporting value in Europe but often sold for aquaria and ponds. Of value in North America for angling and pond culture. **Conservation Status** Alien.

Brown Bullhead *Ameiurus nebulosus*

Size 20–30 cm; maximum 45 cm; maximum weight about 2.5 kg. **Distinctive features** broad flat head with 8 barbels; barbs on posterior pectoral spines well developed; anal fin with 21–24 rays. **Distribution** native to eastern North America, but introduced to Europe (e.g. France), and successfully established in rich lakes and slow-flowing rivers. **Reproduction** June–July, in warm shallow water, spawning in a nest selected by both parents. Both parents guard the nest and the subsequent shoals of young fish. 6,000–13,000 eggs per female. Matures at age 3–4 and lives for about 9 years. **Food** plants and invertebrates, especially large crustaceans. Fish and frogs are also eaten by large adults. **Value** of little commercial or sporting value in Europe, but sold for ponds and aquaria. Important in North America for angling and pond culture. **Conservation Status** Alien.

Channel Catfish *Ictalurus punctatus*

Size 35–50 cm; maximum 120 cm; weight 1–2 kg; maximum 27 kg; world rod record 26.309 kg. **Distinctive features** blue-grey back with bluish silvery sides with black spots. Mouth inferior; 4 pairs of barbels. **Distribution** native to North América in many waters from south Quebec, west to south Alberta and much of the continent south of this. Introduced to many parts of Europe where it has established in Cyprus and Italy. **Reproduction** May–July in nests built by the males under banks, logs, etc. Eggs are guarded by the male until they hatch in 5–10 days at 16–28°C. Growth is rapid, young reaching 10–20 cm after one year and becoming sexually mature at 3–5 years. **Food** bottom invertebrates when young, invertebrates and fish when larger. **Value** sold in Europe for aquaria and ponds but not important commercially. A major commercial and sport fish in North America, where it is widely reared for the table. **Conservation Status** Alien.

European Catfish Family
Siluridae

The Siluridae, a major family of catfishes in the Old World, are found in Europe and Asia – mainly the southern parts. There are 8 genera in the family, but only one of these occurs in Europe. All catfish in this family are scaleless with an elongate body and a very large anal fin, terminating near the caudal fin. There is no adipose fin. Barbels are present on both upper and lower jaws. These fish can grow to an enormous size and are among the largest freshwater fish in the world.

Key to Siluridae in Europe

1	Lower jaw with 2 barbels	*Silurus aristotelis* p166
	Lower jaw with 4 barbels	*Silurus glanis* p167

Aristotle's Catfish *Silurus aristotelis*

Size 1–1.5 m; maximum about 2 m; maximum weight about 150 kg. **Distinctive features** large flat head with two long barbels on upper jaw and 2 short barbels on lower jaw; long anal fin. **Distribution** occurs only in southern Greece in the basin of the Akheloos River, where it is found in slow-flowing water or rich lakes. **Reproduction** June–August, spawning in a nest selected and guarded by the male until the eggs hatch. **Food** mainly zooplankton when very small, then changing to bottom-dwelling invertebrates and later to fish when adult. **Value** of some minor local value commercially and for angling. **Conservation Status** Lower Risk.

Wels Catfish *Silurus glanis*

Size 1–2 m; maximum 5 m – this fish weighed 306 kg; British rod record 19.730 kg. **Distinctive features** large flat head with two long barbels on upper jaw; long anal fin. **Distribution** found in the lower reaches of large rivers and in muddy lakes in most of central and eastern Europe. Occurs also in brackish water in the Baltic and Black Seas. Introduced successfully to a number of other countries (e.g. England). **Reproduction** May–July, spawning in shallow water under thick vegetation where the male fish excavates a depression in which the eggs are laid. These are guarded by the male until they hatch in 2–3 days. The young mature after 4–5 years, and may live up to 15 years. 136,000–467,000 eggs per female. **Food** invertebrates when young, vertebrates (especially fish but also amphibians and waterfowl) when older. **Value** of considerable commercial importance in eastern Europe where it is caught in nets, in traps and on large baited hooks. It is also produced in a few fish farms. Eggs may be used as caviar. Prized as a large angling species in some areas, where it is caught on set lines baited with fish or frogs. **Conservation Status** Lower Risk.

Pike Family
Esocidae

The Esocidae, or pike, occur in lakes and slow-flowing rivers only in the Northern Hemisphere – in temperate Europe, Asia and North America. All have an elongate head, a massive mouth and large teeth. Although widespread, only 1 genus exists, with 5 species – three in North America, one in Siberia and the Northern Pike, which occurs throughout the holarctic region.

Only 1 species occurs in Europe.

Northern Pike *Esox lucius*

Size 30–120 cm; maximum 150 cm; weight up to 34 kg; Scottish rod record 22.085 kg (1947). **Distinctive features** head and snout large, with enormous mouth and teeth. Slim streamlined body with dorsal and anal fins well back towards tail; 110–130 lateral scales. **Distribution** common in standing and slow-flowing waters in temperate Europe, Asia and North America. **Reproduction** February–May among weed. The adhesive eggs hatch after 10–15 days and larvae attach to weed for a short time. They mature in 3–4 years and may live to 25 years. 16,600–79,700 eggs per female. **Food** invertebrates at first, but soon fish and other vertebrates are eaten. Cannibalistic and can exist on its own kind where no other species is present. **Value** of commercial importance and a popular table fish in central and other parts of Europe. A major sport species. **Conservation Status** Lower Risk.

Mudminnow Family
Umbridae

The Umbridae, or mudminnows, are related to the Esocidae and also occur only in the north temperate areas of Europe, Asia and North America, where their distribution is very disjointed. They are found mainly in weedy ponds and streams. Mudminnows are small pugnacious fish with round fins and dull colours. They are very tolerant of low oxygen and low temperatures. There are 3 genera (*Dallia*, *Novumbra* and *Umbra*) and 5 species. Only 2 species occur in Europe, one of them introduced from North America.

Key to Umbridae in Europe

1 More than 14 dorsal rays; 33–35 lateral scales; body blotched;
   lower jaw light   *Umbra krameri* p169

   Less than 14 dorsal rays; 35–37 lateral scales; body striped; jaw dark   *Umbra pygmaea* p169

European Mudminnow Umbra krameri

Size 5–9 cm; maximum 11.5 cm; maximum weight 27 g. **Distinctive features** more than 14 dorsal rays; 33–35 lateral scales. **Distribution** native to Europe, mainly the Danube basin, although introduced elsewhere. Has declined significantly in recent years due to drainage and pollution of habitat. **Reproduction** February–April in nests which the female guards. Young mature in 2 years and live up to 7 years. 1,580–2,710 eggs per female. **Food** invertebrates, especially insect larvae and crustaceans. **Value** of no commercial or sporting importance, but kept in aquaria – though an aggressive species. An endangered species – protected in several countries. **Conservation Status** Endangered.

Eastern Mudminnow Umbra pygmaea

Size 4–8 cm; maximum 10 cm. **Distinctive features** less than 14 dorsal rays. **Distribution** native to eastern temperate North America, but introduced to Europe, probably released by aquarists. **Reproduction** March–April, among weeds in shallow water. Eggs hatch in 5–10 days and fish mature in 2 years. May live for up to 5 years. **Food** invertebrates, especially insect larvae and crustaceans. **Value** of no commercial or sporting value, but often kept in aquaria – though an aggressive species. **Conservation Status** Alien.

169

Smelt Family
Osmeridae

The Osmeridae or smelts occur all round the Northern Hemisphere where various species are marine, anadromous or freshwater in habit. There are 6 genera with 10 species in the Atlantic, Arctic or Pacific Oceans and their basins. They are relatively small fishes which sometimes shoal in enormous numbers when they form an important food source to many predators. All have adipose fins and are small, silvery, slender fish with elongate laterally compressed bodies and large mouths with well developed teeth. Most smelts are popular food species and have a characteristic odour, similar to fresh cucumber.

Only 1 species occurs in fresh water in Europe.

European Smelt *Osmerus eperlanus*

Size 10–20 cm; maximum 31 cm; fish 16 cm long weigh about 32 gm. **Distinctive features** small silvery fish with an adipose fin; ventral fins each with an axillary process; large mouth and teeth; lateral line incomplete. **Distribution** round the coast of north-western Europe and in the estuaries and lower reaches of unpolluted rivers. Some populations in lakes in Scandinavia are purely freshwater. Many stocks have declined due to river barriers and pollution. **Reproduction** March–May, spawning among weeds in rivers above estuaries, or at the edges of lakes. Eggs hatch in 20–35 days and the young move down to the sea, maturing after 2 years. They live up to 8 years. 10,000–40,000 eggs per female. **Food** invertebrates (especially crustaceans) when young, invertebrates and fish when older. **Value** of commercial value in some rivers where they are caught by nets and traps. Introduced as a forage fish to lakes in Scandinavia. **Conservation Status** Vulnerable.

Whitefish Family
Coregonidae

The Coregonidae, or whitefishes, or ciscoes, are closely related to the Salmonidae and, like them, possess a well-developed adipose fin. The elongate body is covered by relatively large cycloid scales, but unlike the Salmonidae, obvious spotting or colour patterning of any kind is absent. Most forms are silvery brown, grey or grey-blue. Teeth are usually either very small or absent.

Members of the family are restricted to the Northern Hemisphere and are found in cool, unpolluted waters in northern Europe, Asia and North America. Their identification presents enormous difficulties, and while it is accepted that there are probably only 3 genera (*Stenodus*, *Prosopium* and *Coregonus*), the number of species within these (especially *Coregonus*) is a matter of considerable dispute. There are probably between 20 and 30 species altogether in the world today. The problem in assigning *Coregonus* to particular species lies in the apparent plasticity of form which these fish are able to exhibit in different habitats. It appears that many of the characteristics which are normally used to diagnose species (shape, size, growth rate, and numbers of scales and gill rakers) are very variable and directly influenced by the environment. Much of the proof for this variability has been obtained from experimental work, moving certain races from one environment to another or rearing different species together in experimental ponds. More recently, however, biochemical techniques involving the use of electrophoresis and DNA analysis techniques appear to have solved many taxonomic problems. However, there is available within the group a large number of scientific names, many outdated, some of which refer to single populations occurring in a lake or river system. Kottelat (1997) considers that many of the older scientific names are legitimate and suggests that there are at least 45 species of *Coregonus* in Europe (excluding Russia). The present account does not accept his arguments.

The Coregonidae are mainly a freshwater group, although some anadromous populations occur in northern seas. They are common in many lakes, especially large, cool unpolluted waters in highland or northern areas. Many forms are entirely pelagic, and some benthic in habit, while other populations are benthic in habit at one time of the year (usually winter) and pelagic at another (summer). They often move around in large shoals and feed on benthic invertebrates and zooplankton. Spawning takes place during

Complete gill arch and raker systems from Coregonus lavaretus (a) and C. albula (b)

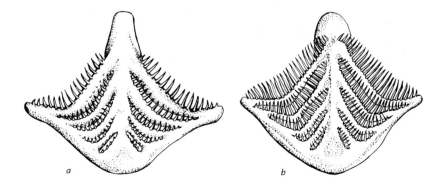

a

b

Key to Coregonidae in Europe

1	Mouth large, oblique and terminal; lower jaw prominent and articulating with cranium behind posterior margin of eye; small teeth present on jaws; 19–26 gill rakers	*Stenodus leucichthys* p176
	Mouth small, lower jaw not prominent and articulating before, or under, eye; jaws toothless or occasionally with minute teeth	Go to **2**
2	Mouth terminal or superior; maxillary reaching behind anterior margin of eye	Go to **3**
	Mouth inferior; maxillary not reaching behind anterior margin of eye	Go to **4**
3	Mouth superior, upper jaw shorter than lower; 36–52 gill rakers	*Coregonus albula* p173
	Mouth terminal, upper jaw equal to lower; 41–68 gill rakers	Go to **7**
4	Well-developed snout; distance between anterior of eye and tip of snout more than twice eye diameter; 35–44 gill rakers	*Coregonus oxyrinchus* p175
	Snout not well developed; distance between anterior of eye and tip of snout less than twice eye diameter	Go to **5**
5	Maxillary wide and short, width usually more than half length, length usually less than 22 per cent head length; snout humped before eyes; 20–29 gill rakers	*Coregonus nasus* p174
	Maxillary narrow and long, width usually less than half length, length usually more than 22 per cent head length; snout not humped before eyes	Go to **6**
6	Length of lower jaw exceeding minimum body depth; 17–48 gill rakers	*Coregonus lavaretus* p174
	Length of lower jaw less than minimum body depth; 16–29 gill rakers	*Coregonus pidschian* p176
7	41–48 gill rakers; 64–67 vertebrae	*Coregonus autumnalis* p173
	49–68 gill rakers; 56–62 vertebrae	*Coregonus peled* p175

the colder months of the year (often under ice), and huge numbers congregate in the spawning areas at this time. The eggs are laid in open water, usually over a stony or gravelly bottom into which the eggs sink and are protected, and hatch in the spring. The young are pelagic for at least the first year of life.

Although rarely of value as sport fish (the Sheefish and a few others are exceptions), the Coregonidae form the basis of some of the most important commercial and subsistence freshwater fisheries in the world, and are a major source of fish flesh in parts of Europe (Russia, Scandinavia and the alpine countries), northern Asia and North America (especially the Great Lakes and other large waters in Canada). Fish are caught in trap nets, gill nets and by seining, and are sometimes dried or smoked. Adult fish have a pleasant, slightly oily, flesh which has often a characteristic odour, likened to that of fresh cucumbers (cf. Grayling and Smelt). In recent years the populations in many of the most important waters have collapsed under fishing pressure, pollution and other man-made impacts, and fisheries which yielded up to 40 million fish annually now give nothing. In Russia, coregonid species have been introduced to large reservoirs, and form the basis of important fisheries there.

Some 8 principal species may be recognised in Europe, several of which are similar to forms found in North America.

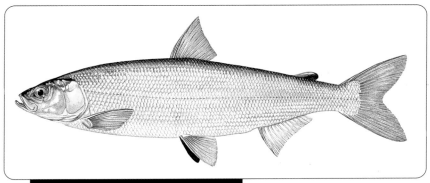

Vendace *Coregonus albula*

Also known as Cisco. **Size** 15–25 cm; maximum 33 cm; fish with a mean weight of 18 cm weigh 60 g. **Distinctive features** mouth superior; 36–52 gill rakers; 75–88 lateral scales; usually small. **Distribution** found in many parts of northern and central Europe, usually in deep cold lakes. Some populations in the Baltic are anadromous. **Reproduction** October–December; eggs are laid over and sink among gravel or stones at depths of 2–3 m or more. Eggs hatch after 100–120 days and the young move into deeper water. Adult mature after 2–3 years and may live up to 10 years. 1,700–4,800 eggs per female. **Food** invertebrates, especially planktonic crustaceans. **Value** of no sporting importance, but important commercially in many areas (e.g. Finland and Russia), fish being caught in nets and traps during their spawning migration. Sometimes smoked. **Conservation Status** Lower Risk.

Arctic Cisco *Coregonus autumnalis*

Size 30–35 cm, 450 g; maximum 38 cm, exceptionally 45 cm (1.3 kg) in Canada and even 50 cm (2 kg) in Siberia. **Distinctive features** mouth terminal, 74–86 lateral scales, 35–51 gill rakers. **Distribution** occurs in western Europe only in large lakes in Ireland (e.g. Lough Neagh); also found in coastal waters and the lower parts of arctic rivers in northern Eurasia and North America. **Reproduction** October–December over areas of gravel and stones. Eggs hatch in March and the young reach 20 cm in 2 years and are adult at 3–4 years. They normally live for 5–7 years (maximum 9–10). 2,000–8,000 eggs per female (exceptionally 90,000 in very large females). **Food** zooplankton when young, zooplankton and bottom invertebrates when older. **Value** fished commercially in Ireland using seine and gill nets. Important commercially in both Siberia and North America. **Conservation Status** Endangered.

Common Whitefish *Coregonus lavaretus*

Size 15–40 cm; maximum 57 cm; maximum weight about 2.8 kg; former British rod record 652 g (now a protected species). **Distinctive features** mouth small, toothless and inferior; 17–48 gill rakers; 84–100 lateral scales. A very variable species forming local races in different lakes, many of them named as subspecies. **Distribution** north-western and central Europe in large cold lakes, and in the Baltic area where it migrates into rivers to spawn. **Reproduction** October–January, over gravel; timing may be influenced by moon phases. Eggs hatch in about 100 days. Young mature in 3–4 years and live up to 10 years. 1,000–28,000 eggs per female. **Food** invertebrates and planktonic crustaceans when young, but both bottom-dwelling and planktonic forms later. **Value** of commercial importance in various countries, and caught mainly by gill nets. Occasionally angled for. **Conservation Status** Lower Risk.

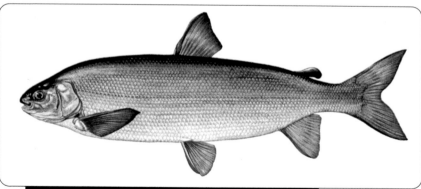

Broad Whitefish *Coregonus nasus*

Size 35–60 cm; maximum 86 cm – this fish weighed 10 kg. **Distinctive features** mouth small, toothless and inferior; maxillary wide and short; snout humped. **Distribution** a holarctic species found in Europe, Asia and North America, mainly in lakes and the lower reaches of rivers associated with the Arctic Ocean. **Reproduction** October–November in rivers, the young migrating back to lakes where they mature after 4–5 years; they live for up to 10 years. 14,000–29,000 eggs per female. **Food** invertebrates, especially insects and molluscs. **Value** caught commercially and in subsistence fisheries in a few areas where it is gill netted and sometimes dried and smoked. **Conservation Status** Lower Risk.

Houting *Coregonus oxyrinchus*

Size 25–40 cm; maximum 50 cm; maximum weight about 2 kg. **Distinctive features** mouth small, toothless and inferior; well-developed snout; 35–44 gill rakers. Some believe this fish is merely a subspecies of *Coregonus lavaretus*. **Distribution** northern Europe, especially the Baltic Sea, where large populations formerly occurred in brackish water, only running into fresh water to spawn. **Reproduction** October–December, in the lower reaches of rivers. After hatching, the young migrate to the sea, where they mature in 3–4 years. **Food** invertebrates, planktonic crustaceans when young, planktonic and benthic forms later. **Value** formerly locally important in some areas and caught during the autumn spawning migration. Many populations are now extinct but recently a conservation programme in Denmark has proved successful in restoring some stocks. **Conservation Status** Endangered.

Peled *Coregonus peled*

Size 30–40 cm; maximum 50 cm; maximum weight about 5 kg. **Distinctive features** mouth terminal and toothless; 49–68 gill rakers. **Distribution** found in much of northern Europe, particularly in lakes in arctic areas of Siberia, Finland and Sweden. **Reproduction** September–November, in shallow water in both lakes and rivers – often migrating considerable distances to the spawning grounds. 29,000–105,000 eggs per female. **Food** invertebrates, especially planktonic crustaceans. **Value** a very important commercial species, netted in large numbers in many Russian waters. **Conservation Status** Lower Risk.

Arctic Whitefish *Coregonus pidschian*

Size 20–30 cm; maximum 50 cm; a 48 cm fish weighs about 1.25 kg. **Distinctive features** mouth inferior; maxillary narrow and long; 16–29 gill rakers. Some workers equate this species with the *Coregonus clupeaformis* complex in North America. **Distribution** arctic areas of Europe in rivers, lakes and some coastal areas of the Baltic. **Reproduction** October–December, eggs hatching in spring. Adults mature at 2–4 years and live up to 12 years. 8,000–50,000 eggs per female. **Food** invertebrates, especially crustaceans. **Value** of considerable commercial importance in many areas. **Conservation Status** Lower Risk.

Sheefish *Stenodus leucichthys*

Also known as Inconnu. **Size** 50–80 cm; maximum 130 cm; maximum weight about 35 kg. The largest coregonid fish. **Distinctive features** prominent lower jaw with small teeth; usually large. **Distribution** occurs in Europe and North America in lakes and rivers whose basins are linked to the Arctic Ocean. A population also occurs in the northern Caspian Sea where it is found in some coastal areas and associated rivers. **Reproduction** October–November, usually in the lower or middle reaches of large rivers. Adults mature at 8–10 years of age and may live up to 20 years. 130,000–420,000 eggs per female. **Food** invertebrates when young, but mainly fish (e.g. *Coregonus*) when adult. **Value** of some sporting significance in a few parts of North America and an important commercial species in both Canada and Russia. Most fish are caught in nets during the spawning migration and many are stored after smoking or drying. **Conservation Status** Lower Risk.

Salmon Family
Salmonidae

The family Salmonidae includes several groups of fish known commonly as salmon, trout and charr. None of this important family is entirely marine, all its members being either anadromous or purely freshwater. Members are native to Europe, north-west Africa, northern Asia and North America, but have been introduced to many other parts of the world, including South America, India, Australia and New Zealand. Much of the original stock for these introductions came from Europe.

Salmonidae possess an adipose fin, and the body is covered with well-developed cycloid scales. Spots of various kinds are a common feature of coloration, and the young stages of many species are characterised by dark parr marks along the sides. Teeth are usually present and well developed in some species, and there are often marked differences between the sexes at spawning time. The taxonomy of most groups within the family has been the subject of much debate and argument over more than 200 years, but most taxonomists have favoured groupings of many of the original 'species' into fewer accepted species. However, Kottelat (1997) considers that many of the older scientific names are legitimate and suggests that there are at least 28 species of *Salmo* and 23 species of *Salvelinus* in Europe (excluding Russia). The present account does not accept his arguments.

The biology of all species in the family presents many common features. All of them spawn in fresh, usually running, water, but sometimes on clean gravelly areas of lake bottoms. No matter which part of the world is concerned, they are found only in cool water (i.e. only highland areas in the tropics), and spawn during the coldest months of the year. Spawning usually takes place in pairs: after a sexual display, the female digs out a nest, or redd, from the gravelly substrate, and in this they spawn. After the eggs have been laid they are covered over with gravel by the female.

Cycloid scale from a trout, showing annual rings

Members of the family Salmonidae are prized by anglers and commercial fishermen in many countries, and good populations are a valuable natural resource. The Atlantic Salmon is a major resource in many countries, both for angling but also, more recently, for farming, and the present European output of farmed Atlantic Salmon (mainly from Norway, Scotland and Ireland) is about 300,000 tonnes – many times that of wild caught fish. The North Atlantic Salmon Conservation Organisation (NASCO) is an intergovernmental group which seeks to conserve this species in all parts of its range.

The Pacific salmon (*Oncorhynchus*) are of major importance in western North America and north-east Asia. The 5 species occurring here form the basis of many huge fisheries which have been operating for more than 100 years. Competition for these stocks among Pacific nations (especially the USSR, Japan, Canada and the USA) became so great that it was necessary to set up an international commission to regulate

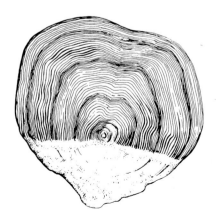

Salmon Family Salmonidae

Key to Salmonidae in Europe

1	Postorbital bones reaching preopercular; 12–19 branched rays in anal fin	*Go to* **2**
	Postorbital bones not reaching preopercular; 8–12 branched rays in anal fin	*Go to* **4**
2	No distinct black spots on back and caudal fin; 16–26 gill rakers	***Oncorhynchus keta*** p181
	Distinct black spots on back and caudal fin	*Go to* **3**
3	Black spots on back and caudal fin large; lateral scales 147–205; 16–22 gill rakers	***Oncorhynchus gorbuscha*** p180
	Black spots on back and caudal fin small; lateral scales 112–148; 18–25 gill rakers	***Oncorhynchus kisutch*** p182
4	Vomer elongate, always toothed posteriorly in young specimens	*Go to* **5**
	Vomer short and wide, toothless posteriorly even in young specimens	*Go to* **9**
5	More than 130 scales along lateral line; no red spots on body, but a broad pink band along sides; numerous black spots on body and fins, including adipose and tail fins	***Oncorhynchus mykiss*** p183
	Less than 130 scales along lateral line; the body may be completely silver, but normally has many black and some red spots; black spots on the adipose and tail fins are ill defined or absent; no broad pink band along side	*Go to* **6**
6	Less than 110 scales along lateral line	***Salmothymus obtusirostris*** p186
	More than 110 scales along lateral line	*Go to* **7**
7	Head of vomer toothless, the shaft poorly toothed with deciduous teeth; 10–12 rays in dorsal fin; 10–13 scales between adipose fin and lateral line	***Salmo salar*** p184
	Head of vomer toothed, the shaft also well toothed with persistent teeth; 8–10 rays in dorsal fin; 13–16 scales between adipose fin and lateral line	*Go to* **8**
8	Body with dark spots and perhaps dark parr marks, but no irregular dark marbling	***Salmo trutta*** p185
	Body with irregular dark marbling over most parts	***Salmo marmoratus*** p183
9	Well-developed space between vomerine and palatine teeth; no dark spots on body	*Go to* **10**
	Vomerine and palatine teeth forming a continuous horseshoe-shaped patch; dark spots present on body	***Hucho hucho*** p179
10	Caudal fin deeply forked; body and fins covered with small, oval light spots, never orange or red on body; more than 90 pyloric caeca	***Salvelinus namaycush*** p188
	Caudal fin not, or only slightly, forked; body and fins with pink or red spots and usually dark lines and marks; less than 75 pyloric caeca	*Go to* **11**
11	Hyoid teeth present; premaxillary not toothed on right side; back uniformly coloured with pale spots; black stripe absent on anal fin	***Salvelinus alpinus*** p187
	Hyoid teeth absent; premaxillary toothed on right side; back strongly vermiculated; black stripes present on anal fin	***Salvelinus fontinalis*** p188

the fishery and carry out research on the biology of the species concerned. The results of this research have led to many valuable discoveries about *Oncorhynchus* and other fish, including information on fish navigation (both in the sea and in fresh water), the physiology of adjustment to fresh or salt water, and factors controlling the abundance of stocks. Major canning industries have been established on many rivers in which there are runs of Pacific salmon, and both canned and smoked fish are exported to all parts of the world. Attempts to introduce these fish to other parts of the world have been less successful than with trout, but populations of four species are now apparently established in northern Europe.

There are 12 species (6 of them introduced from North America) in Europe.

Huchen

Taimen

Huchen *Hucho hucho*

Size 50–100 cm; maximum over 100 kg; a 120 cm fish weighs 21 kg; largest fish recorded 105 kg (1943). **Distinctive features** 10 or less branched rays in anal fin; vomer toothless; dark spots on body; 9–19 gill rakers on 1st arch; 107–194 lateral scales. **Distribution** an exclusively freshwater species, occurring as the Huchen *Hucho hucho hucho* in the Danube river system and as the larger Taimen *Hucho hucho taimen* in the Volga and large rivers further east. **Reproduction** March–May, spawning among gravel in flowing water. The eggs hatch in about 35 days and adults mature in 4–5 years. Adults may reach an age of 20 years (Huchen) or 50 years (Taimen), but it has been suggested that some may live even to 100 years or more. Females with a mean weight of 5 kg have about 5,000 eggs. **Food** invertebrates and fish when young, but mainly fish (especially *Chondrostoma*) when larger. **Value** the largest salmon in the world – a highly prized sport species whose numbers have diminished alarmingly in recent years. Farms have been established for stocking. **Conservation Status** Endangered.

Pink Salmon *Oncorhynchus gorbuscha*

Size 40–50 cm; maximum 64 cm; fish with a mean length of 50 cm weigh about 1.7 kg. **Distinctive features** 10 or more branched rays in anal fin; 177–240 scales along lateral line. Also known as Humpback Salmon from the shape of adults. **Distribution** native to the north Pacific and associated large rivers in North America and Asia. Occurs sporadically in northern European seas and appears to have established in some rivers as a result of introduction to the White Sea area. **Reproduction** September–October, in nests among gravel in running water. The eggs hatch after 100–120 days and the young migrate very quickly downstream to the sea, where they grow fast, maturing after 2 years. 1,090–1,629 eggs per female. **Food** mainly invertebrates (especially larger crustaceans) and fish. **Value** a major commercial and sporting species in some north Pacific coastal areas, and fresh waters. **Conservation Status** Alien.

Chum Salmon *Oncorhynchus keta*

Size 45–90 cm; maximum 100 cm; fish with a mean length of 75 cm weigh 4.9 kg. **Distinctive features** 10 or more branched rays in anal fin; 150–160 scales along lateral line. **Distribution** native to the north Pacific and associated large rivers in North America and Asia. Occurs sporadically in northern European seas and some rivers as a result of regular introductions to the White Sea area where it appears to have established in a few areas. **Reproduction** August–December in areas of coarse gravel where a reasonable current is present. The eggs are laid in nests excavated by the female, and hatch in 90–130 days. The fry remain in the river for several months, but have normally migrated to the sea by the end of their first year. The adults mature after 3–4 years and migrate back to their parent rivers. 2,008–4,388 eggs per female. **Food** invertebrates when young, large invertebrates (mainly crustaceans) and fish when adult. **Value** a major commercial and sporting species in the north Pacific, both in the sea and in fresh water. **Conservation Status** Alien.

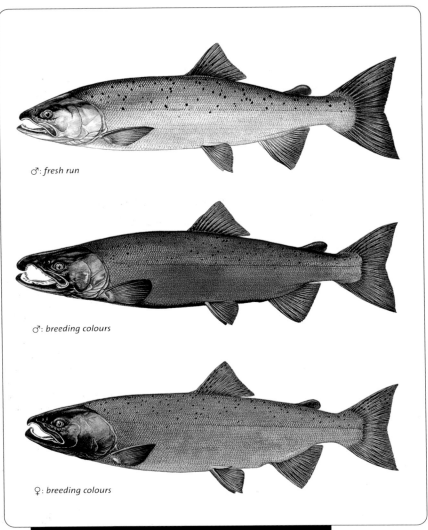

♂: *fresh run*

♂: *breeding colours*

♀: *breeding colours*

Coho Salmon *Oncorhynchus kisutch*

Size 45–60 cm; maximum 100 cm; fish with a mean length of 55 cm weigh 5 kg; British rod record 0.682 kg (1977); world rod record 14.062 kg. **Distinctive features** rough striations on caudal fin rays; 121–148 lateral scales. **Distribution** native to the northern Pacific and coastal streams from Sakhalin north to Alaska and south again to Monterey, but introduced elsewhere, notably the Great Lakes; introduced also to several European countries but established only in France. **Reproduction** October–March, in nests among gravel in running water. Eggs hatch in 35–50 days; some fry migrate almost immediately to the sea but most remain in fresh water for one year (exceptionally two in the north). Growth in the sea is rapid, some fish maturing at 2 years, most at 3–4 years. They die after spawning. 1,400–6,000 eggs per female. **Food** mainly aquatic insects in fresh water when young; invertebrates and fish in the sea during the marine phase. **Value** of relatively little value in Europe, but of major importance as a commercial and sport species in North America. **Conservation Status** Alien.

Rainbow Trout *Oncorhynchus mykiss*

Size 25–45 cm; maximum 70 cm and 20 kg; world rod record 19.108 kg. **Distinctive features** 10 or less branched rays in anal fin; caudal and adipose fins heavily spotted; broad pink band along sides. Formerly known as *Salmo gairdneri*. **Distribution** native to the Pacific coastal basins of North America and Asia where both migratory (Steelhead Trout) and purely freshwater populations occur. Has been introduced widely all over Europe, although it is only in a few places that self-maintaining populations have developed. **Reproduction** October–March, the eggs being laid in nests dug out of gravel in running water. The eggs hatch after 100–150 days and the young gradually move into larger rivers and lakes. 1,000–5,000 eggs per female. **Food** invertebrates when young, invertebrates and fish when adult. **Value** of major importance as a commercial species in many fish farms in Europe and North America and as a stocked sporting species in rivers and lakes. **Conservation Status** Alien.

Marbled Trout *Salmo marmoratus*

Size 35–50 cm; maximum 100 cm, with a maximum weight of 20 kg. **Distinctive features** characteristic marbled pattern on body with brownish-olive coloration interwinding with a light silver-grey. **Distribution** rivers in northern Italy, the Adriatic river basins of Slovenia and Croatia, Montenegro and Albania. **Reproduction** December–January, spawning among clean gravels. Egg development takes 430–470 day-degrees and the young reach 8 cm after 1 year, 15 cm after 2 years. Sexually mature after 3–4 years. Females have about 1,800 eggs per kg of body weight. **Food** young fish feed largely on bottom invertebrates but adults are almost exclusively piscivorous. **Value** a valuable and highly prized angling species. Unfortunately, many populations have disappeared or declined – often due to competition with and introgression by introduced *Salmo trutta*, of which some believe *Salmo marmoratus* is merely a subspecies. **Conservation Status** Endangered.

fry

parr

smolt

♂

♀

Atlantic Salmon *Salmo salar*

Size 40–100 cm; maximum 120 cm; British rod record 29.028 kg; world rod record 35.891 kg. **Distinctive features** 10 or less branched rays in anal fin; vomer toothed posteriorly but not anteriorly. **Distribution** anadromous native species widely found in the Atlantic areas of northern Europe and eastern North America in clear stony rivers, streams and accessible lakes. **Reproduction** October–January. Eggs, laid in redds among gravel in running water, hatch after 70–160 days (depending on water temperature) into alevins, which change to fry, then to parr. After 2–6 years these become silver coloured and migrate to the sea as smolts. Major growth occurs in the sea (fish returning within 1 year are called grilse; later than 1 year salmon). A 70 cm female carries 5,000–6,000 eggs. **Food** invertebrates when small, invertebrates and fish when larger. **Value** a major commercial species, netted off Atlantic coasts and estuaries and off Greenland. An important sport species angled in rivers and lakes – fly fishing, spinning and trolling all successful in various waters. A major farmed species in several countries (e.g. Norway, Scotland and Ireland). **Conservation Status** Vulnerable.

Freshwater form (Brown Trout)

Sea run form (Sea Trout)

Brown Trout *Salmo trutta*

Size 15–50 cm; maximum 70 cm; British rod record 8.745 kg; world rod record 16.301 kg. **Distinctive features** 10 or less branched rays in anal fin; vomer toothed anteriorly and posteriorly; less than 130 scales along lateral line. Numerous subspecies have been described. **Distribution** occurs all over Europe in many running and standing waters. Migratory populations occur in rivers all along the European coast from Spain northwards. **Reproduction** September–October in small rivers and streams, where eggs are laid in a nest (redd) which the female cuts out in the gravel. The eggs hatch in approximately 150 days and the fry spend a year or more in the nursery stream before moving down into a larger river or lake (or the sea, in the case of the migratory Sea Trout). Adults mature after 3–5 years and many live up to 20 years. The mean number of eggs is about 5,400 per female. **Food** invertebrates when young, invertebrates and some fish (especially in the case of sea trout) when adult. **Value** an important commercial species in some areas, and possibly the most important single sport species over Europe as a whole. Widely introduced to many parts of the world, including North America, where it is well established in some areas. **Conservation Status** Lower Risk.

Adriatic Salmon *Salmothymus obtusirostris*

Size 25–40 cm; maximum 50 cm; maximum weight about 5 kg. **Distinctive features** 101–103 lateral scales; head short. **Distribution** found only in waters in Dalmatia this species is considered by many ichthyologists to be derived from isolated populations of Atlantic Salmon. Several local races are recognised. **Reproduction** October–December among gravel in running water. **Food** invertebrates when young, invertebrates and fish when older. **Value** of little commercial or sporting interest. A related fish *Salmothymus ohridanus*, whose taxonomic status is uncertain and may be a subspecies of *Salmothymus obtusirostris*, occurs in Lake Ohrid (Albania/Macedonia). **Conservation Status** Critically Endangered.

Arctic Charr *Salvelinus alpinus*

Size 15–40 cm; maximum 88 cm; world rod record 14.770 kg. **Distinctive features** 10 or less branched rays in anal fin; vomer toothless posteriorly; body uniformly coloured with pale spots. Many different morphs occur, sometimes in the same lake where there may be migratory, normal and dwarf populations living together. **Distribution** a holarctic species found in many northern catchments in the Northern Hemisphere in clear cool lakes. In Arctic seas anadromous stocks are found, maturing in the sea and entering rivers to spawn. These fish are larger than the freshwater forms, some of which are dwarf races. **Reproduction** October–March, spawning among gravel in both lakes and rivers. After hatching the young move into lakes or the sea, maturing in 3–6 years. 560–7,300 eggs per female. **Food** mainly invertebrates, especially planktonic crustaceans. Large fish in some populations may be piscivorous (even cannibalistic). **Value** netted in Arctic areas during the spawning migration and in many lakes. Important subsistence fisheries occur among inuit people in North America. A sporting species in some lakes. **Conservation Status** Lower Risk.

Salmon Family Salmonidae

American Brook Charr *Salvelinus fontinalis*

Size 20–35 cm; maximum 50 cm; world rod record 6.577 kg. **Distinctive features** 10 or less branched rays in anal fin; vomer toothless; body with a marked vermiculate pattern with dark and light areas interwoven. **Distribution** native to western North America, but introduced to many parts of Europe where it has become established. **Reproduction** October–March among gravel in running waters. The eggs hatch in spring and the young spend about 2 years in nursery areas before moving into larger rivers or lakes. Adults mature at 2–3 years of age. 100–5,000 eggs per female. **Food** mainly invertebrates, but fish are eaten by large adults. **Value** a major sport species in North America, and popular in some places in Europe. **Conservation status** Alien.

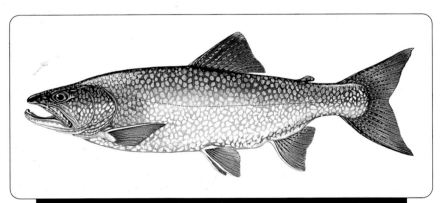

American Lake Charr *Salvelinus namaycush*

Size 30–50 cm; maximum 126 cm – this fish weighed 46.3 kg; world rod record 29.484 kg. **Distinctive features** 10 or less branched rays in anal fin; vomer toothless; body covered with light yellowish spots. **Distribution** widespread in northern United States and Canada, including the Great Lakes. Introduced to Europe where populations are established in large deep lakes in Sweden and Switzerland. **Reproduction** September–November, spawning over stones and gravel. Eggs hatch in 100–150 days and the young move offshore soon afterwards. Adults mature after 6–8 years, but may live up to 25 years. An 80 cm female has about 18,000 eggs. **Food** invertebrates when young, but other fish (especially *Coregonus*) as size increases. **Value** an important commercial and sport species in North America, but of little importance in Europe. **Conservation Status** Alien.

Grayling Family
Thymallidae

The Thymallidae, or graylings, are found only in the northern hemisphere (Europe, Asia and North America) where 4 species in 1 genus occur. They are related to both the Salmonidae and the Coregonidae. All live in swift-flowing streams and cold lakes, occasionally in estuaries. The name of this genus refers to the flesh which is supposed to smell like the herb thyme (cf. Common Whitefish and Smelt). The high dorsal fin is typical of the family.

Only 1 species occurs in Europe, but the Arctic Grayling *Thymallus arcticus* is found in northern Siberian rivers and in northern North America.

European Grayling *Thymallus thymallus*

Size 25–35 cm; maximum 50 cm; maximum weight 4.675 kg; Scottish rod record 1.389 kg (1994). **Distinctive features** adipose and large dorsal fins. The spot pattern is specific to individual fish. **Distribution** found in clean cool rivers (occasionally lakes) in parts of Europe, especially northern areas. Sometimes in estuaries. **Reproduction** March–May, eggs laid in nests in gravel hatch in 20–30 days and the young mature at 3–4 years. They live up to 15 years. Females about 45 cm long carry about 10,000 eggs. **Food** mainly invertebrates, especially insects. Larger individuals may eat small fish. **Value** of some commercial importance to fisheries especially in Russia. A popular sport fish in some countries, such as Britain, where it is caught on worms or artificial flies. **Conservation Status** Lower Risk.

Cod Family
Gadidae

The Gadidae, or cods, are mainly marine fish, found in cool waters in the Northern Hemisphere and to a lesser extent the Southern Hemisphere. Many species are intertidal, others oceanic, but the majority occur at moderate depths on the continental shelf. There are about 60 species, but few enter fresh water. Cods have wide heads with large jaws and numerous fine teeth. Some have barbels on the upper lip and many have one or more barbels on the chin. Characteristically, the pelvic fins are placed well forward, in front of the level of the pectoral fin base. They are major commercial fish (e.g. Cod *Gadus morhua*, Haddock *Melanogrammus aeglefinus* and Whiting *Merlangius merlangus*), especially in the north Atlantic where many are also of sporting value.

Only 1 species occurs in fresh water in Europe.

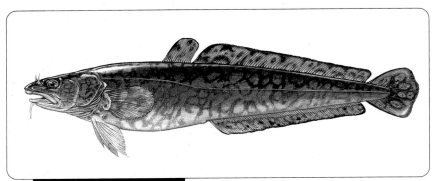

Burbot *Lota lota*

Size 30–50 cm; maximum 120 cm; maximum weight about 32 kg. **Distinctive features** elongate body with 2 dorsal fins, the first shorter than the second; broad head with 1 long barbel on lower jaw, and 1 shorter barbel at each nostril. **Distribution** lakes and rivers throughout northern areas of Europe, Asia and North America. Has declined in some parts of its range (e.g. it is now extinct in Great Britain). **Reproduction** December–March, over stones and gravel in rivers and lakes. Eggs hatch in 40–50 days and the young mature after 3–4 years. Can live for up to 25 years. 33,082–3,063,000 eggs per female. **Food** when young, invertebrates (especially crustaceans and insect larvae); when older, fish. **Value** of some commercial value, caught by means of nets, traps and baited hooks, but of little sporting value. **Conservation Status** Vulnerable.

Grey Mullet Family
Mugilidae

The Mugilidae, or grey mullets, are common fish in most oceans of the world, particularly in shallow inshore waters. Many species enter brackish and fresh water, but usually only estuaries, lagoons and the lower reaches of rivers, and for short periods – sometimes only a few hours at a time, on a tidal basis. There are several genera with a total of about 100 species found in tropical and temperate waters.

The mullets are elongate but sturdy fish, whose bodies are only slightly compressed laterally. The mouth is terminal and large, but teeth are either very small or absent. In some species adipose eyelids are well developed. The function of these is believed to be that of protection and streamlining (see page 192). The body is covered by large, usually cycloid, scales which extend on to the head. There is no lateral line. The gill rakers are long and slender. There are two, well-separated dorsal fins, the anterior fin being short and supported by 1–5 strong spines. The second dorsal fin is longer, and supported by softer branched rays. The pectoral fins are set rather high and immediately behind the gill covers.

Mullets are fast-swimming shoaling fish which often come into shallow waters in large numbers to feed. Much of their food consists of filamentous algae, but invertebrates are also eaten in some numbers. The gut is remarkably elongate, with a powerful muscular gizzard situated anteriorly, which crushes and breaks up food which is subsequently digested posteriorly in the intestine.

Spawning takes place in the sea, usually during spring in inshore waters. Relatively little is known about their breeding biology, however. Mullet fry commonly abound along some shores and enter streams in these areas. The adults run into larger rivers from time to time.

Although a widespread family, the importance of mullets to man can vary greatly from place to place. They are angled for along some coasts, usually by casting lines from the shore, but sometimes from boats. Commercial fishermen mostly use traps and seine nets, and catches in some areas are large – up to 20 million kg (19,683 tonnes) per annum in the Mediterranean area. In southern Russia the roe as well as the flesh of the fish is important commercially.

Six species occur around European shores and enter fresh waters from time to time.

Key to Mugilidae in Europe

1	Preorbital bone posteriorly not reaching the level of the corner of the mouth; maxillary entirely concealed beneath preorbital bone; adipose eyelid well developed, covering eye to the pupil	*Mugil cephalus* p195
	Preorbital bone posteriorly reaching beyond level of the corner of the mouth; maxillary projecting from under preorbital bone at corner of mouth; adipose eyelid weakly developed, never reaching the pupil	*Go to 2*
2	Upper lip very thick: its depth not less than 1/10th of head length and more than 1/2 of eye diameter; branches of the lower jaw not covered with scales	*Go to 3*
	Upper lip less thick: its depth less than 1/10th of head length and not more than 1/2 of eye diameter. Branches of the lower jaw scaled	*Go to 4*
3	Upper lip smooth, its thickness about equal to the diameter of the eye	*Oedalechilus labeo* p195
	Upper lip with two rows of small warts, its thickness less than the diameter of eye	*Chelon labrosus* p193
4	Some dorsal scales with several (2–5) canals; upper side of head scaled to the anterior nostrils; scales on the snout terminating in numerous rows of small scales; no elongate lobule above the base of the pectoral fin; several golden spots on gill covers	*Liza saliens* p194
	Dorsal scales with single canals	*Go to 5*
5	Upper side of head scaled to the nostrils or even anteriorly; posterior end of preorbital rounded or truncated vertically (but not obliquely); elongate lobule above base of pectoral fin	*Liza ramada* p194
	Scales on upper side of head not reaching anterior of nostrils, and terminating in a single series of small scales; posterior end of preorbital obliquely truncated; no elongate lobule above base of pectoral fin; 1 golden spot behind each gill cover, and 1 behind each eye	*Liza aurata* p193

Characteristic features of Mugilidae:
a = dorsal view of Sharpnose Mullet
showing sensory canals; b = side view of
Striped Mullet showing adipose eyelid

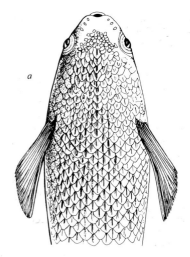

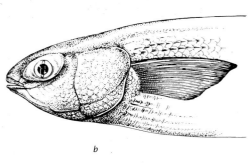

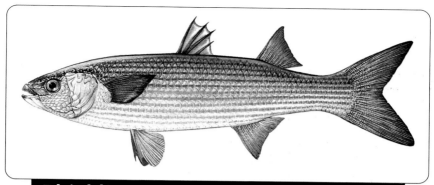

Thicklipped Grey Mullet *Chelon labrosus*

Size 30–50 cm; maximum 90 cm; British rod record 6.428 kg (1979). **Distinctive features** body covered with large cycloid scales which extend on to head; thick adipose eyelid; 2 dorsal fins, the first with only 4 spiny rays; upper lip very thick and with 2 rows of small warts; 45–46 lateral scales. **Distribution** found in Atlantic coastal areas south of Norway and in the Mediterranean and Black Seas, occurring in fresh water in the lower reaches of some rivers. **Reproduction** June–August, in the sea. The young mature after 2–4 years. **Food** filamentous algae and other bottom plants, together with associated invertebrates. **Value** of considerable commercial value (along with other mullets) in net and trap fisheries. Considered a sport species in some areas. **Conservation Status** Lower Risk.

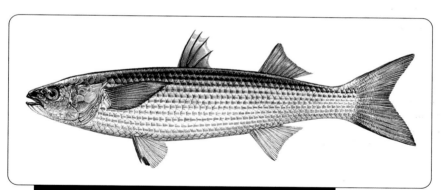

Golden Grey Mullet *Liza aurata*

Size 20–35 cm; maximum 50 cm; British rod record 1.368 kg (1994). **Distinctive features** body covered with large cycloid scales which extend on to head; operculum and cheek with golden patches; 2 dorsal fins, the first with only 4 spiny rays; no elongate lobule above base of pectoral fin; golden spot on gill cover and behind each eye. **Distribution** found all round the Atlantic coastal areas of southern Europe, including the Mediterranean and Black Seas. It has been successfully introduced to the Caspian Sea. Enters the lower reaches of a number of rivers (e.g. Dnieper). **Reproduction** August–September, spawning in the sea. The young mature after 3–5 years and may live up to 10 years. 1,200,000–2,100,000 eggs per female. **Food** mainly filamentous algae and other bottom plant material, but also invertebrates (e.g. molluscs) at times. **Value** of considerable commercial value in net and trap fisheries, especially in the Mediterranean and Black Seas. Of little sporting significance. **Conservation Status** Lower Risk.

193

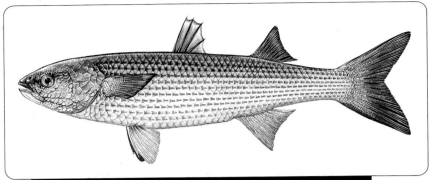

Thinlipped Grey Mullet *Liza ramada*

Size 25–40 cm; maximum 60 cm; British rod record 3.175 kg (1991). **Distinctive features** upper lip with only one row of small tubercles; body covered with large cycloid scales, with one groove, which extend on to head; 2 dorsal fins, the first with only 4 spiny rays; no golden spots on gill covers. **Distribution** found round all parts of the European coast (including the Mediterranean and Black Seas) except the extreme north. Enters the lower reaches of large rivers and some coastal lakes and lagoons. **Reproduction** August–September, spawning in the sea. The young mature after 3–5 years. **Food** mostly filamentous algae and associated invertebrates (e.g. molluscs). **Value** of considerable commercial and some sporting value in various parts of its range. **Conservation Status** Lower Risk.

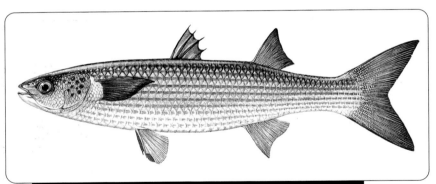

Sharpnose Grey Mullet *Liza saliens*

Size 20–30 cm; maximum 40 cm. **Distinctive features** body covered with large cycloid scales, with two or more grooves, which extend on to head; 2 dorsal fins, the first with only 4 spiny rays; several golden spots on gill covers. **Distribution** Atlantic coast of southern Europe and North Africa, Mediterranean and Black Seas. Has been introduced to the Caspian Sea. Found in fresh water in the lower reaches of a number of rivers (e.g. Guadalquivir) and in coastal lagoons. **Reproduction** May–June, in shallow seas. **Food** mainly benthic algae and other plant material including associated invertebrates (e.g. molluscs). **Value** of some commercial significance in net and trap fisheries in the Mediterranean and Black Seas. **Conservation Status** Lower Risk.

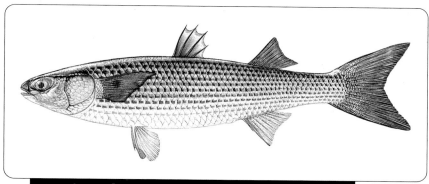

Striped Grey Mullet Mugil cephalus

Size 30–60 cm; maximum 70 cm; maximum weight about 6 kg. **Distinctive features** body covered with large cycloid scales extending on to head; 2 dorsal fins, the first with only 4 spiny rays; adipose eyelid well developed, covering eye to pupil. **Distribution** found widely all round Atlantic coastal areas of southern Europe (including the Mediterranean and Black Seas) and North America. It has been introduced to the Caspian Sea. Enters fresh waters via large estuaries and coastal lakes and has a wide tolerance of salinity. **Reproduction** June–August, spawning in the sea. The young mature after 6–8 years, 5,000,000–7,200,000 eggs per female. **Food** mainly filamentous algae and other plant material, but also various invertebrates. **Value** of considerable commercial value, including fish farming; large catches are made by nets and traps in several countries. Angled for in estuarine waters in some places. **Conservation Status** Lower Risk.

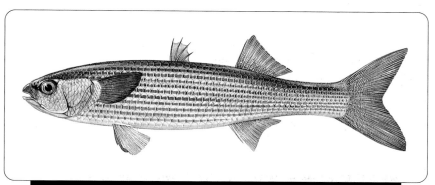

Boxlipped Grey Mullet Oedalechilus labeo

Size 15–20 cm; maximum 25 cm. **Distinctive features** body with large cycloid scales which extend on to head; 2 dorsal fins, the first with only 4 spiny rays; upper lip smooth and equal in thickness to the eye diameter; eye mostly covered by adipose eyelid. **Distribution** found in coastal areas of Europe from France round to the northern Mediterranean. Occurs in fresh water in the lower reaches of large rivers. **Reproduction** July–September, spawning in the sea. **Food** fine plant material and benthic invertebrates, especially molluscs and crustaceans. **Value** of considerable commercial value to coastal fisheries in France. Of little sporting significance. **Conservation Status** Lower Risk.

Silverside Family
Atherinidae

The Atherinidae, also known as silversides or sandsmelts, are found around the world except in high latitudes. They are plankton feeders and occur in the sea, brackish waters and various fresh waters. There are 50 genera and about 170 species. Most are small (less than 15 cm) although the largest species grow to more than 50 cm. All are silvery translucent fish, usually with a silvery lateral stripe.

Only 5 species occur in Europe – they are mainly marine but may enter fresh water.

Key to Atherinidae in Europe

1	1st dorsal fin well back, almost above anus and much smaller than 2nd dorsal fin. Large species (up to 70 cm), found only in Lake Nemi near Rome	*Odonthestes bonariensis* p198
	1st dorsal fin well forward of anus and similar in size to 2nd dorsal fin. Small species (less than 25 cm) found elsewhere	*Go to* **2**
2	Less than 52 lateral scales	*Go to* **3**
	More than 51 lateral scales	*Go to* **4**
3	42–48 lateral scales; 40–47 vertebrae	*Atherina boyeri* p196
	48–51 lateral scales; 48–52 vertebrae	*Atherina mochon* p197
4	Less than 12 branched rays in 2nd dorsal fin; 59–65 lateral scales	*Atherina hepsetus* p197
	More than 12 branched rays in 2nd dorsal fin; 52–57 lateral scales	*Atherina presbyter* p198

Bigscale Silverside *Atherina boyeri*

Size 7–9 cm; maximum 13 cm; British rod record 8 g (1980). **Distinctive features** 42–48 lateral scales; 40–47 vertebrae; Anal fin with 11–13 branched rays. **Distribution** around the Mediterranean and Caspian Seas in intertidal areas, lagoons and some fresh waters (e.g. Lake Trichonis). **Reproduction** June–August, among vegetation to which the eggs adhere by means of filaments. **Food** invertebrates, mainly small crustaceans with some worms and molluscs. **Value** of no sporting value and only minor commercial significance to local fisheries. **Conservation Status** Lower Risk.

Common Silverside *Atherina hepsetus*

Size 10–12 cm; maximum 15 cm. **Distinctive features** more than 55 lateral scales; less than 12 branches rays in dorsal fin. **Distribution** along the coasts of the Mediterranean, Black, and Caspian Seas, and in some estuaries and lakes. **Reproduction** June–August, spawning among vegetation to which the eggs adhere by filaments. **Food** invertebrates, mainly crustaceans. **Value** of no sporting value, and only minor commercial significance to local fisheries. **Conservation Status** Lower Risk.

Mediterranean Silverside *Atherina mochon*

Size 8–12 cm; maximum 14 cm. **Distinctive features** less than 55 lateral scales. **Distribution** along the coasts of the Mediterranean, Black and Caspian Seas and in the estuaries and lower reaches of some rivers (Dniester, Bug). **Reproduction** July–August, spawning among vegetation, to which the eggs attach by fine filaments. **Food** benthic invertebrates, especially crustaceans and (in fresh water) insect larvae. **Value** of no sporting and only minor commercial significance to fisheries in the Black and Caspian Seas. **Conservation Status** Lower Risk.

Atlantic Silverside *Atherina presbyter*

Size 10–13 cm; maximum 15 cm; British rod record 72 g (1975). **Distinctive features** more than 55 lateral scales; more than 15 branched rays in dorsal fin. **Distribution** along the coasts of Europe from Denmark to Spain and the western Mediterranean. Enters some estuaries. **Reproduction** April–July, in shallow water among seaweed to which the filamentous eggs attach. **Food** invertebrates, especially crustaceans. **Value** of no sporting, but minor commercial, importance in fisheries along the Mediterranean coast. **Conservation Status** Lower Risk.

Pejerrey Silverside *Odonthestes bonariensis*

Size 40–50 cm; maximum 70 cm – much larger than any native European silversides. **Distinctive features** mouth terminal; 1st dorsal fin almost above anus and much smaller than 2nd dorsal fin. **Distribution** native to South America, but successfully introduced to Lake Nemi, near Rome, to replace *Coregonus* there. **Reproduction** not known in Lake Nemi. **Food** zooplankton, mainly cladocerans and copepods. **Value** of little commercial or sporting value in Europe. **Conservation Status** Alien.

Killifish Family
Fundulidae

The Killifish family Fundulidae includes a large number of species but these are found only in North and Central America and neighbouring islands – except where they have been introduced elsewhere by humans. All representatives have a cylindrical body a flattened head and a somewhat compressed caudal peduncle. They occupy a wide variety of habitat – some are surface living, others bottom living in fresh water, whilst yet others are found in brackish water and even in the sea. The sexes are easily distinguished: the males have larger fins than the females and are always better marked and more brilliantly coloured. An unusual feature of females is that they have a 'sexual sac' which is an external prolongation of the oviduct, surrounding the genital opening and supported by the first ray of the anal fin. Members of the family are popular with aquarists and widely sold in many countries.

No species is native to Europe, but one species occurs here as a result of successful introductions to the Iberian Peninsula.

♂

♀

Mummichog *Fundulus heteroclitus*

Size 8–10 cm; maximum 13 cm **Distinctive features** deep caudal peduncle; mouth superior; 8–12 gill rakers; 32–38 lateral scales. **Distribution** a North American species which occurs in coastal, brackish and some fresh waters from the Gulf of St Lawrence to northern Florida. Now established in Europe on the Iberian Peninsula as a result of introductions. **Reproduction** June–July in pairs, with courtship and mating followed by deposition of eggs on the bottom. About 400–500 eggs per female. Eggs hatch in 10–20 days, depending on temperature. **Food** omnivorous feeding on vegetation, small invertebrates, including amphipods and molluscs, and small fish and fish eggs. **Value** used extensively for experimental work (embryology and physiology) in the past, also as live bait by anglers. Of no commercial importance. **Conservation Status** Alien.

Valencia Family
Valenciidae

The Valenciidae is a monogeneric family with only two species in Europe. They are related to the toothcarps (Cyprinodontidae) but differ in that *Valencia* has unicuspid conical teeth whereas *Aphanias* has tricuspid teeth. The body is elongate and somewhat compressed laterally. All members of the family are small fish which live in small standing waters, often near the sea.

Key to Valenciidae in Europe

1 Anal fin with 12–14 rays; caudal fin with 4–7 dark bars. found only in Spain   *Valencia hispanica* p200

   Anal fin with 11–13 rays; caudal fin with 2–4 dark bars; found only in Greece and southern Albania   *Valencia letourneuxi* p201

♀

♂

Spanish Valencia *Valencia hispanica*

Size 5–7 cm; maximum 8 cm. **Distinctive features** mouth terminal; dorsal fin anterior to anal fin, the latter with 12–14 rays; dark spot above base of pectoral fin. **Distribution** found only in coastal areas of south-east Spain, occurring in a great variety of waters, including small pools and ditches. Inhabits both fresh and brackish waters. **Reproduction** April–July, spawning among weed in shallow water. The eggs hatch in 12–15 days and the young mature after less than one year. About 200–250 eggs per female. **Food** invertebrates, especially crustaceans and insect larvae. **Value** of no real commercial or sporting value, although sometimes kept in aquaria. **Conservation Status** Endangered.

Greek Valencia *Valencia letourneuxi*

Size 5–7 cm; maximum 8 cm. **Distinctive features** anal fin with 11–13 rays; caudal fin with 2–4 dark bars. **Distribution** found only in the eastern Mediterranean, from southern Albania to the Peloponnese. Occurs along this coastal area in lagoons and freshwater springs. **Reproduction** March–June among vegetation in shallow water. **Food** small invertebrates, especially crustaceans and insect larvae. **Value** of no commercial or sporting value but sometimes kept by aquarists. **Conservation Status** Endangered.

Toothcarp Family
Cyprinodontidae

The Cyprinodontidae, or toothcarps, are found mainly in fresh water, but sometimes in brackish and occasionally marine situations. They occur only in warm water and are found in southern Europe, Africa, south-east Asia, South America and south-eastern U.S.A. They are all small fish, often with a characteristically upturned mouth and well-developed, but small, tricuspid teeth. Many species are very popular as aquarium fish.

There are a large number of genera with many species, but only 2 of these occur in Europe.

Key to Cyprinodontidae in Europe

1 9–10 rays in dorsal fin; male caudal fin bears several vertical bands; female sides mottled green-brown	*Aphanius iberus* p203
10–13 rays in dorsal fin; male caudal fin bears 1 vertical band; female sides grey with faint darker vertical bands	*Aphanius fasciatus* p202

♂

♀

Fasciated Toothcarp *Aphanius fasciatus*

Size 4–5 cm; maximum 6 cm. **Distinctive features** small fish with superior mouth; dorsal fin, with 10–13 rays, positioned well back and almost immediately above anal fin, which has 9–13 rays. **Distribution** found along the coastal regions of much of the eastern Mediterranean area, occurring in a variety of fresh and brackish waters including small weedy ponds and ditches. **Reproduction** April–August, among vegetation. The eggs hatch in 10–15 days and the young mature within one year. **Food** invertebrates, especially crustaceans and insect larvae. **Value** of no commercial or sporting significance, although sometimes kept in aquaria. **Conservation Status** Lower Risk.

Iberian Toothcarp *Aphanius iberus*

Size 3–4 cm; maximum 5 cm. **Distinctive features** small fish with superior mouth; dorsal fin, with 9–10 rays, placed well back and almost immediately above anal fin, which has 9–10 rays. **Distribution** found only in the coastal areas of south-eastern Spain in a variety of both fresh and brackish waters including small ditches and pools. Occurs also in Algeria. **Reproduction** April–August, spawning among vegetation in shallow water. The eggs hatch in 10–15 days and the young mature within one year. About 200 eggs per female. **Food** invertebrates, mainly crustaceans and insect larvae. **Value** of no commercial or sporting value, although sometimes kept in aquaria. **Conservation Status** Vulnerable.

Livebearer Family
Poeciliidae

The Poeciliidae is a family of livebearing fish found mainly in the warm temperate and tropical areas of North, Central and South America where more than 25 genera, with some 140 species, are known to occur. All are viviparous and in the males the anal fin is modified, its 3rd, 4th and 5th rays being elongated to form a copulatory organ called a gonopodium. Although very commonly kept in tropical aquaria in Europe, none are native here and only 1 species has been successfully introduced to the wild. However, one other species, the Guppy *Poecilia reticulata*, not dealt with in this book, can maintain itself where there are heated effluents – in England (where it is now extinct) and in the Netherlands.

Mosquito Fish *Gambusia affinis*

Size 3–5 cm; maximum 6 cm. **Distinctive features** mouth superior; dorsal fin placed well back, above or posterior to anal fin; male with anal fin modified to form penis. There are two subspecies, *Gambusia affinis affinis* and *Gambusia affinis holbrooki*: both have been introduced to Europe at different times. **Distribution** native to south-eastern areas of North America, this species has been introduced to several parts of Europe in an attempt to control mosquitoes. It is now widely established in southern Europe and is found in a variety of fresh and brackish waters including small weedy ditches and pools. **Reproduction** mainly April–August. This is a livebearing species in which the females are fertilised internally by the males and give birth to some 50 young about 30 days afterwards. Several broods may be produced each year and the young mature within one year. **Food** invertebrates, especially crustaceans and insect larvae. **Value** of importance in mosquito control, especially in malarial areas, but a hazard to some native species. Commonly kept as an aquarium fish. **Conservation Status** Alien.

Stickleback Family
Gasterosteidae

The Gasterosteidae, popularly known as sticklebacks, occur in a great variety of aquatic habitats in the Northern Hemisphere in Europe, Asia and North America. They are found in the sea, in estuaries and brackish waters, and in all kinds of fresh waters except where the flow is extremely fast. Although widespread, there are relatively few species, probably 8, distributed in 5 genera.

Sticklebacks are all small laterally compressed fish with well-developed dorsal and pelvic fin spines – these being characteristic of the family. There are really two dorsal fins, the first being represented by a series of spines. The pelvic fins have virtually been replaced by the spines. The mouth and teeth are small, but the latter are important for crushing and chewing. The body has no proper scales and is either naked or covered to a variable extent by bony plates.

Although small, sticklebacks are particularly aggressive fish and feed on a great variety of invertebrates and fish fry. The European species have a very characteristic behaviour at spawning time, when each male establishes a territory in which it builds a nest. After spawning, the females are chased away and the eggs and subsequent fry are guarded carefully by the males. Further details of the spawning behaviour of the Three-spined Stickleback are described on page 23.

Four species are found in fresh waters in Europe. A fifth species, *Spinachia spinachia* (L.), the Sea Stickleback, is found almost always in marine (occasionally estuarine) conditions.

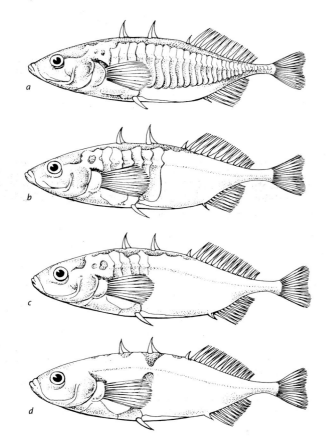

Variation in armouring in different races of Gasterosteus aculeatus:
a. trachurus type (mainly brackish);
b. and c. intermediate types;
d. leiurus type (mainly freshwater)

Key to Gasterosteidae in Europe

1	2–3 spines anterior to dorsal fin	*Gasterosteus aculeatus* p207
	4–12 spines anterior to dorsal fin	*Go to* **2**
2	4–6 spines anterior to dorsal fin; body with obvious dark spotted or mottled patterning	*Pungitius hellenicus* p208
	7–12 spines anterior to dorsal fin. Body with no obvious dark patterning	*Go to* **3**
3	Caudal peduncle usually with a well-developed lateral keel; body naked	*Pungitius pungitius* p209
	Caudal peduncle smooth, with no lateral keel; body with lateral scutes	*Pungitius platygaster* p208

♂

♀

Three-spined Stickleback
Gasterosteus aculeatus

Size 4–8 cm; maximum 11 cm. **Distinctive features** 3 strong spines anterior to dorsal fin; no scales on body but this may be protected by a variable number of bony plates. At spawning time the male develops a bright red throat and is iridescent green elsewhere. In some parts of Northwest Scotland, spineless forms occur. **Distribution** found in many parts of Europe, especially areas which are not too far from the sea. Occurs in a wide variety of waters from the sea to rivers and lakes of all kinds. **Reproduction** March–June, when the male builds a nest of fibrous material and induces one or more females to lay in it. The eggs hatch in 5–20 days and they, and the young, are guarded by the male. The young mature after 1–2 years and rarely live beyond 4 years. 90–450 eggs per female. **Food** invertebrates (mainly worms, crustaceans and insect larvae) and sometimes small fish. **Value** formerly used in parts of Europe for the production of fish meal, it is now of little commercial or sporting significance. It is commonly kept in aquaria, however, and is often used for teaching purposes. **Conservation Status** Lower Risk.

Greek Stickleback *Pungitius hellenicus*

Size 3–4 cm; maximum 5 cm. **Distinctive features** 2–5 dorsal spines, variable mottled or spotted colour pattern on sides. **Distribution** found only in Greece in karstic springs within the catchment of the Sperchois River. All are clean, cool (never more than 12°C), well-oxygenated waters with rich vegetation. **Reproduction** unknown. **Food** unknown, but presumably small invertebrates. **Value** of no commercial value. An endangered species now protected by Greek law. **Conservation Status** Endangered.

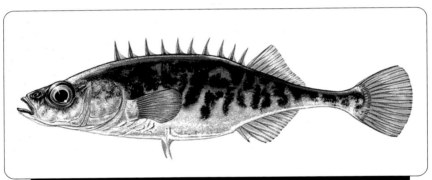

Ukrainian Stickleback *Pungitius platygaster*

Size 4–6 cm; maximum 7 cm. **Distinctive features** 8 or 9 strong spines anterior to dorsal fin; no scales present, but sides of body covered with inconspicuous bony scutes; ventral spine distinctly serrated. **Distribution** found in a variety of waters (both fresh and brackish) within basins linked to the north of the Black, Caspian and Aral Seas. **Reproduction** April–May, spawning in nests built by the males in shallow water, in well-vegetated areas. The male protects the eggs and young for some time. **Food** invertebrates, especially crustaceans and insect larvae. **Value** of no commercial or sporting value. **Conservation Status** Lower Risk.

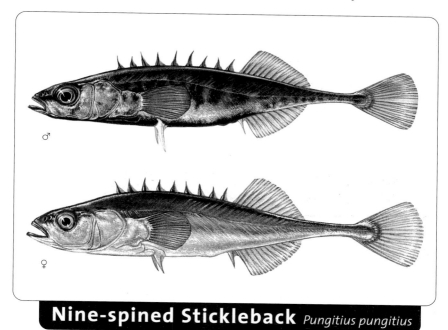

♂

♀

Nine-spined Stickleback *Pungitius pungitius*

Size 5–7 cm; maximum 9 cm. **Distinctive features** 7–12 – although usually 9 – stiff spines anterior to dorsal fin; body without scales. At spawning time the male develops a black throat and becomes very dark elsewhere. **Distribution** found in most parts of northern Europe, Asia and North America which are not too far from the sea. Found in both brackish and fresh waters. **Reproduction** April–July, the male builds a nest of fine plant material among vegetation, and induces one or more females to lay in it. The eggs hatch after 10–20 days; both they and the fry are guarded by the male. The young mature after 1 year and live for only 2–3 years at most. **Food** invertebrates, especially crustaceans and insect larvae. **Value** of no commercial or sporting value. Sometimes kept in aquaria. **Conservation Status** Lower Risk.

Pipefish Family
Syngnathidae

The Syngnathidae, or pipefish, are common in most seas and oceans, particularly among seaweed in shallow coastal areas. There are many genera and species, but only a few enter fresh water and even these are common only in estuaries, coastal lagoons and lower reaches of some rivers.

All pipefish have a very elongate body with or without pectoral and caudal fins. The dorsal fin is usually moderately well developed, but ventral fins are absent. There are no scales, but the body is covered with, and protected by, a series of bony rings which form a strong outer skeleton. The snout is very elongate, with a very small terminal mouth and two nasal openings on each side. There are no teeth. The gill openings are very small, situated behind the upper edges of the gill covers.

Most members of the family are rather poor swimmers and live in protected places among weed and rocks. In such areas their elongate shape and mottled colouring give them considerable camouflage protection. Because of the minute size of the mouth, only small animals can be eaten – these are usually crustacean zooplankton and fish fry.

A very characteristic feature of the group is the fact that the male has a marsupial pouch on the underside of the tail and abdomen. It is formed by two folds of skin which are developed on either side and meet in the mid-line. In a few members (e.g. *Nerophis*) the pouch is absent and the eggs are attached directly to the abdomen. The inside of the pouch is lined with soft skin. During spawning the eggs are fertilised and placed in the brood pouch of the male, where they remain for the full period of incubation. Soon after hatching, the young, which, apart from size are very similar to the adults, are released and immediately start to fend for themselves.

Four species are found in fresh waters in some areas of Europe.

Syngathidae courtship behaviour: eggs being transferred from the female to the pouch of the male

Male pouch opened to show eggs

Key to Syngnathidae in Europe

1	Pectoral fins absent in adults	***Nerophis ophidion*** p211
	Pectoral fins present in adults	*Go to* **2**
2	Both halves of the pectoral ring mobile below and not fused; usually no median ventral bony plate under pectoral ring; proboscis strongly compressed laterally	***Syphonostoma typhle*** p212
	Both halves of the pectoral ring fused below and immobile; median ventral bony plate present; proboscis subcylindrical and elongate	*Go to* **3**
3	Dorsal fins occupying 7–9 rings, and with 29–42 rays	***Syngnathus abaster*** p211
	Dorsal fins occupying 11–13 rings, and with 41–62 rays	***Syngnathus acus*** p212

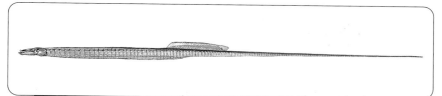

Straightnose Pipefish *Nerophis ophidion*

Size 15–25 cm; maximum 30 cm. **Distinctive features** no pectoral fins in fish longer than 10 cm. **Distribution** European coast from northern Norway to the Mediterranean and Black Seas. Enters the lower reaches of rivers (e.g. Dniester, Dnieper). **Reproduction** May–August, eggs being fertilized by the male after transfer to his pouch. Young mature after 1 year and live for 3–4 years. 200–300 eggs per female. **Food** small invertebrates (mainly crustaceans) and fish larvae. **Value** of no commercial or sporting significance. **Conservation Status** Lower Risk.

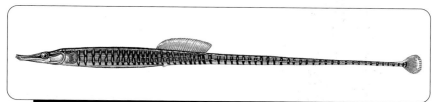

Shore Pipefish *Syngnathus abaster*

Size 15–18 cm; maximum 22 cm. **Distinctive features** dorsal fins occupying 7–9 rings and with 29–42 rays; proboscis elongate and cylindrical. **Distribution** found among vegetation throughout the Mediterranean, Black and Caspian Seas. Ascends the lower reaches of rivers entering these seas (e.g. Dniester, Volga) and occurs also in lakes on their plains. **Reproduction** May–August; after spawning eggs and fry are retained in pouches on males. Eggs hatch in 20–25 days and young mature after 1 year. **Food** small invertebrates (mainly crustaceans) and fish larvae. **Value** of no commercial or sporting significance. **Conservation Status** Lower Risk.

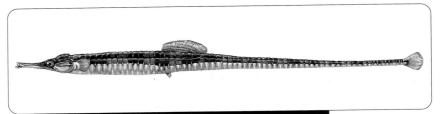

Greater Pipefish *Syngnathus acus*

Size 20–35 cm; maximum 46 cm. **Distinctive features** dorsal fin occupying 11–13 rings and with 41–62 rays; proboscis elongate and cylindrical; colour brown with dark green vertical bars. **Distribution** along Atlantic coasts from western Norway to Spain and throughout Mediterranean. Occurs in estuaries. **Reproduction** June–July, eggs are transferred from the female to the male's brood pouch where they are fertilised, and hatch in 30–35 days at a length of 22–35 mm. 200–400 eggs per female. **Food** small invertebrates (mainly crustaceans and fish larvae). **Value** of no commercial or sporting significance. **Conservation Status** Lower Risk.

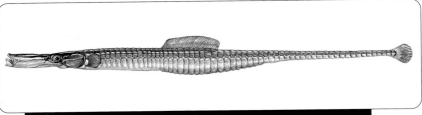

Deepnose Pipefish *Syphonostoma typhle*

Size 20–30 cm; maximum 37 cm. **Distinctive features** proboscis compressed laterally; body pale greenish brown; head with small dark spots. **Distribution** European coasts from Norway to the Mediterranean and Black Seas; among sea grass and algae. Enters estuaries and lower reaches of rivers. **Reproduction** March–August, eggs being transferred from the female to the male brood pouch by an elongate tube. After fertilisation, eggs remain in the pouch until they hatch after 25–30 days. Young mature in 1 year and live 3–4 years. 150–200 eggs per female. **Food** small invertebrates (mainly crustaceans) and fish larvae. **Value** of no commercial or sporting significance. **Conservation Status** Lower Risk.

Sculpin Family
Cottidae

The Cottidae, also known as bullheads or sculpins, are mainly small bottom-living fish which occur in Europe, Asia and North America. They are mostly marine fish and found in cool temperate waters in the Northern hemisphere and over 300 species have been described. Typically, they are stout-bodied, large headed fish with spines on the gill covers and broad pectoral fins.

Only 3 species occur in fresh water in Europe.

Key to Cottidae in Europe

1 Branchiostegal membrane free of isthmus, forming a fold; 4 yellow growths on head	*Triglopsis quadricornis* p214
Branchiostegal membrane attached to isthmus and not forming a free fold; no growths on head	Go to **2**
2 Lateral line extending to caudal fin; inner unpigmented pelvic ray more than 1/2 length of longest ray	*Cottus gobio* p213
Lateral line ending below 2nd dorsal fin; inner pigmented pelvic ray less than 1/2 length of longest ray	*Cottus poecilopus* p214

Common Bullhead *Cottus gobio*

Size 10–15 cm; maximum 18 cm. **Distinctive features** no scales; lateral line with 30–35 pores and ending at caudal fin. *Cottus petiti*, found only in a small catchment in France, is probably a subspecies of *Cottus gobio*. **Distribution** in stony streams and some oligotrophic lakes in most of Europe except the extreme north and south. **Reproduction** March–May, spawning in nests under stones guarded by the male. The pale yellow eggs, laid in clumps, hatch in 20–25 days and young mature in 2 years; rarely living longer than 6 years. About 100 eggs per female. **Food** mainly invertebrates, especially insect larvae, but also fish eggs and fry. **Value** of no commercial or sporting value. **Conservation Status** Lower Risk.

Alpine Bullhead *Cottus poecilopus*

Size 8–10 cm; maximum 13 cm. **Distinctive features** no scales; lateral line with 20–25 pores and ending below 2nd dorsal fin. **Distribution** in stony streams, rivers and some lakes in north and central Europe and northern Asia. **Reproduction** February–April, spawning in nests under stones, where the eggs are laid in clumps. **Food** invertebrates, especially crustaceans and insect larvae. **Value** of no commercial or sporting significance. **Conservation Status** Lower Risk.

Fourhorn Sculpin *Triglopsis quadricornis*

Size 10–25 cm; maximum 60 cm. **Distinctive features** 2 pairs of spongy growths on head; 4 preopercular spines; numerous small tubercles above lateral line. **Distribution** in the sea, and in brackish water along Arctic coasts of Europe, Asia and North America. Occurs as isolated populations in large lakes (e.g. Malarsee, Lake Superior), living often at great depths on soft mud. **Reproduction** December–January, spawning among stones in areas guarded by the males. **Food** benthic invertebrates, especially crustaceans. **Value** of some value to local fisheries. Of no sporting significance. **Conservation Status** Lower Risk.

Bass Family
Moronidae

The Moronidae, or sea basses, as many of them are known, occur in most seas around the world – both temperate and tropical. Although many of the 300 or so species are marine, a few occur in brackish or in fresh waters, especially when they are young. They are characteristically well-built, slim but deep-bodied, fish with spiny anterior dorsal fins and ctenoid scales. Many form the basis of important commercial fisheries in different parts of the world.

Only 2 species enter fresh water in Europe.

Key to Moronidae in Europe

1   Scales on interobital space cycloid; teeth on head of vomer only; no spots on adults
Dicentrarchus labrax p215

Scales on interobital space ctenoid; teeth over whole of vomer; spots present on adults; usually forming longitudinal series
Dicentrarchus punctatus p216

Sea Bass *Dicentrarchus labrax*

Size 40–70 cm; maximum 100 cm; maximum weight about 12 kg; British rod record 8.877 kg (1987). **Distinctive features** 2 equal dorsal fins, which just meet each other; scales between eyes cycloid; 52–74 scales along lateral line; large dark mark on gill cover, but none on body. **Distribution** found all round European coasts (including the Mediterranean and Black Seas) except in the extreme north. Enters the lower reaches of large rivers. **Reproduction** March–June, in the sea off shore, laying planktonic eggs. Large females may have up to 2,000,000 eggs. Mature at 3–4 years and may live to 30 years. **Food** when young, invertebrates (especially molluscs and crustaceans) and some fish; when adult mainly fish (e.g. herring). **Value** of minor commercial importance in some countries. An important sporting species, offering exciting angling in some estuaries and coastal waters. **Conservation Status** Lower Risk.

Spotted Bass *Dicentrarchus punctatus*

Size 25–40 cm; maximum 100 cm. **Distinctive features** 2 equal dorsal fins which just meet each other; scales between the eyes ctenoid; 57–65 lateral scales; body with numerous dark spots, usually forming several longitudinal rows. **Distribution** found round the European coast from northern France to southern Italy. Absent from the eastern Mediterranean. Enters the lower reaches of rivers and their lagoons. **Reproduction** May–August, spawning in estuaries and along coasts. In fresh or brackish water the eggs sink, whereas in sea water they float. **Food** invertebrates (especially molluscs and crustaceans) and fish of various kinds. **Value** of some commercial and sporting significance in the western Mediterranean. **Conservation Status** Lower Risk.

Sunfish Family
Centrarchidae

The Centrarchidae, or sunfish, are native originally to a variety of waters, particularly rich well-weeded lakes and slow-flowing rivers in eastern North America. The present distribution of the family, however, is much wider than this, as various species have been successfully introduced to other parts of the world, including western North America and Europe. The family is a small but important one, including 30 species in 10 genera.

Most sunfish are small to medium-sized, laterally flattened fish with a well-developed dorsal fin, the anterior half of which has very spiny rays. The spiny portion and the soft-rayed portion of the dorsal fin are separated to a varying extent, but are always closer together than in most of the perch family. The other fins are also well-developed, particularly the anal fin, which is usually very similar in size to the posterior part of the dorsal fin. The mouth and eyes are large, and bands of small teeth are found on various parts of the mouth including the tongue.

The members of this family include several highly coloured and very attractive species which exhibit a number of interesting features of behaviour. The smaller species are popular with aquarists and pond keepers, and the larger species with anglers. The Smallmouth Bass and Largemouth Bass in particular are major sporting species, and are an important element in the gigantic sport fishery and associated tourist industries in eastern Canada and parts of the United States. The originally important commercial fishery for these fish is now of very little value.

The breeding biology of most species of Centrarchidae is now well documented and many common features are found throughout the family. *Typical centrarchid behaviour; a male fish guards the nest and eggs* Ripe fish usually spawn in late spring or early summer, when the males select territories and dig out nests. These are simply circular depressions which can vary in size from 10–200 cm in diameter, usually on a sandy or gravelly bottom, but often near the protection of logs, rocks or patches of vegetation. Males may return to the same nest in subsequent years, or at least to the same general area of previous nests. On completion of the nest each pair of

Key to Centrarchidae in Europe

1	6 anal spines arising in a scaled groove; body with 7–9 horizontal rows of black spots below lateral line; base of anal fin about 66% of base of dorsal fin	*Ambloplites rupestris* p219
	3 anal spines, not arising in a groove; no horizontal rows of spots below lateral line; base of anal fin about 33–50% of base of dorsal fin	*Go to* **2**
2	More than 55 scales along lateral line; body length more than 3 times greatest depth	*Go to* **3**
	Less than 50 scales along lateral line; body length less than 3 times greatest depth	*Go to* **4**
3	Upper jaw extending back beyond eye; 60–68 scales along lateral line; pelvic fins not joined by membrane	*Micropterus salmoides* p221
	Upper jaw extending back only to middle of eye; 68–78 scales along lateral line; pelvic fins joined by membrane	*Micropterus dolomieu* p221
4	Opercular flap all black, with no colour round edge	*Lepomis auritus* p219
	Opercular flap with black centre, and yellow, orange or red spots or bands round margin	*Go to* **5**
5	Pectoral fins about 33% of body length, pointed at leading edge; gill rakers knobbed; opercular flap short with a prominent red spot posteriorly	*Lepomis gibbosus* p220
	Pectoral fins only about 25% of body length, rounded at leading edge; gill rakers not knobbed; opercular flap with no prominent red spot posteriorly	*Lepomis cyanellus* p220

fish indulges in considerable courtship display which ends in the spawning act. Eggs are laid regularly over a period of hours until the female is spent. The eggs are adhesive, and usually attach themselves to clean stones and gravel near the centre of the nest. During incubation the male guards the nest, fanning the eggs continually during this time. After hatching, the larvae still have considerable yolk sacs, and lie on the bottom of the nest for several days, still protected and cleaned by the male. Eventually they start to leave the nest but are still guarded by the male for several days. After this the family breaks up and the young fend for themselves.

Six species are known to have become established in various waters in Europe, mainly as a result of introductions by anglers and aquarists.

Rock Bass *Ambloplites rupestris*

Size 15–20 cm; maximum 34 cm; maximum weight 1.7 kg. **Distinctive features** anal fin with 6 spines arising in a scaled groove; 7–9 horizontal rows of black spots below lateral line; 39–40 lateral scales. **Distribution** native to eastern central North America, it has been introduced to Europe and has become established in at least one country (England). **Reproduction** May–July, spawning in a nest excavated by the male among sand and gravel. The male guards the eggs (which hatch in 3–4 days) and early fry. Young fish mature in 2–3 years and live to 10 years. 3,000–11,000 eggs per female. **Food** invertebrates (especially crustaceans and insect larvae) and small fish. **Value** of no significance in Europe but a valuable commercial and sport fish in North America. **Conservation Status** Alien.

Redbreast Sunfish *Lepomis auritus*

Size 12–15 cm; maximum 24 cm, weighing 0.79 kg. **Distinctive features** single dorsal fin divided into two parts; opercular flap all black with no colour round edge. **Distribution** native to fresh waters in eastern North America, this species has been introduced and is established in Europe in Italy. **Reproduction** June–July, spawning in nest excavated by the male which also guards the eggs and early fry. **Food** mainly invertebrates (especially insect larvae), but sometimes small fish. **Value** of no importance in Europe and of only minor local sporting significance in North America. **Conservation Status** Alien.

Sunfish Family Centrarchidae

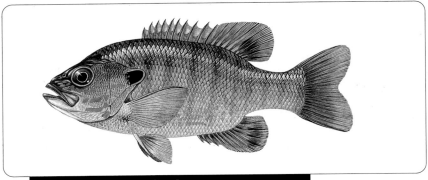

Green Sunfish *Lepomis cyanellus*

Size 10–12 cm; maximum 30 cm – this fish weighed about 1 kg. **Distinctive features** single dorsal fin divided into two parts; gill rakers not knobbed; colour brown to olive with an emerald sheen; no red spot on opercular flap. **Distribution** native to fresh waters in eastern central North America, this species has been introduced to European waters and is established in Germany. **Reproduction** May–August, spawning in a shallow nest excavated by the male in shallow water in areas sheltered by rocks and vegetation. The male guards and fans the eggs and protects the young for a short time. The eggs hatch in 3–5 days and the young mature after 2 years. They may live up to 9 years. **Food** mainly invertebrates (molluscs and insect larvae) but occasionally small fish. **Value** of no commercial or sporting significance in Europe, but an important game species in some parts of the U.S.A.. **Conservation Status** Alien.

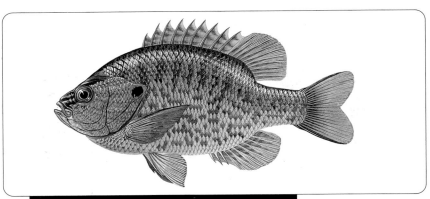

Pumpkinseed *Lepomis gibbosus*

Size 10–15 cm; maximum 22 cm; maximum weight about 300 gm. **Distinctive features** single dorsal fin divided into two parts; gill rakers knobbed; opercular flap short with a prominent red spot posteriorly; 40–47 lateral scales. **Distribution** native to fresh waters in eastern North America, from Canada to Tennessee, this species has been introduced and become established in many parts of Europe. **Reproduction** May–July, spawning in shallow depressions among sand in weed beds. The eggs (which hatch in 3–5 days) and the fry are guarded by the male. The young mature after 2–3 years and may live for 9 years. 600–5,000 eggs per female. **Food** mainly invertebrates (especially crustaceans and insect larvae) but small fish are eaten by adults. **Value** of little significance in Europe, but caught in some numbers in North America both commercially and as a sport fish. **Conservation Status** Alien.

Smallmouth Bass *Micropterus dolomieu*

Size 20–40 cm; maximum 58 cm; maximum weight 6 kg; world rod record 5.415 kg. **Distinctive features** upper jaw extending back to middle of eye; 68–78 lateral scales. **Distribution** native to clear lakes and some rivers in eastern central North America, it has been introduced to Europe where it is now established in several countries (e.g. France). **Reproduction** May–July, spawning in a nest excavated by the male among sand or gravel. The male guards eggs and fry, the former hatching after 4–10 days. The young mature in 3–4 years and may live to 15 years. 5,000–14,000 eggs per female. **Food** invertebrates (mainly crustaceans and insect larvae) when young, large invertebrates and fish when older. **Value** of little significance in Europe, but formerly important commercially in North America and still a prized sport species there. **Conservation Status** Alien.

Largemouth Bass *Micropterus salmoides*

Size 20–40 cm; maximum 83 cm; maximum weight 11 kg; world rod record 10.092 kg. **Distinctive features** upper jaw extending behind eye; 60–68 lateral scales. **Distribution** native to southern Canada and U.SA., it has been introduced to Europe and is established in a number of countries (e.g. Spain). **Reproduction** March–July, spawning in pits dug out in sand and gravel. Eggs are guarded by the male and take 2–5 days to hatch. Young mature after 3–4 years and live up to 15 years. 751–11,457 eggs per female. **Food** invertebrates (especially crustaceans and insect larvae) when young, large invertebrates, fish and frogs when adult. **Value** of relatively little, though increasing, value in Europe. A prized sport fish in North America where it is also caught commercially. **Conservation Status** Alien.

Perch Family
Percidae

The Percidae, or perches, as some members are commonly called, are restricted to the Northern Hemisphere and occur in various waters in Europe, northern Asia and North America. Successful introductions of at least one species (European Perch) have been made in the Southern Hemisphere (Australia). There are 6 genera and about 16 species within the family – 13 of these being found in Europe. All but one of the European species are found in fresh water, but a few can live in estuaries and penetrate into brackish water. One species is largely marine.

Typical members of the family are elongate and rather compressed laterally. The mouth is usually large and carries several rows of teeth, many of them sharp, and some elongated to form canines. There are two dorsal fins, usually distinctly separated, with the anterior fin supported by 6–15 strong spines. The rays of the posterior dorsal fin are soft and branched. The body is covered by well-developed ctenoid scales which make the skin rough to the touch. In many species the scales extend on to the head which is also protected by spiny outgrowths from the opercular and other bones.

Several of the larger Percidae are of importance to sport fishermen, particularly the European Perch, the Pikeperch and other species of *Sander*. In both Europe and North America large numbers of these fish are angled for, using a variety of techniques including spinning and live baiting. These same larger species are important commercially in many countries, for the flesh is white, flaky and delicious when cooked properly. They are caught in traps, (European Perch in particular flock into these in enormous numbers in early summer) and nets. In 1934 the catch of Yellow Perch *Perca flavescens* (closely related to the European Perch) alone from the Canadian side of the Great Lakes was over 25 million kg (24,603 tonnes). Lesser, but still valuable, catches are obtained from European waters.

Ctenoid scale from a Perch, showing annual rings

All Percidae are carnivorous, and the larger species live mainly on other fish. For this reason they are sometimes used to control numbers of prolific herbivorous fish in fish farm ponds and elsewhere. In most species spawning takes place in pairs, trios or sometimes even larger numbers, when there are usually several males to one female. The eggs are laid over vegetation, or stones and gravel. In some species (for instance the European Perch), eggs are laid in long strings among weed; in others they are separate, and at spawning time are broadcast loose and fall down among the gravel and stones. There are 13 species in Europe. All of these are native. The North American Walleye *Sander vitreum* has been introduced in Europe, but it is uncertain whether it has ever become established here.

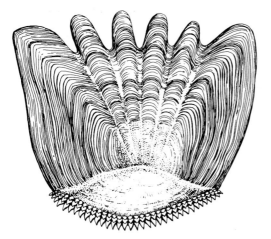

Key to Percidae in Europe

1	Eyes positioned dorsally on head	*Romanichthys valsanicola* p227
	Eyes positioned laterally or dorso-laterally on head	*Go to* **2**
2	Sensory organ cavities weakly developed on sides and top of head	*Go to* **3**
	Large sensory organ cavities laterally and dorsally on head	*Go to* **9**
3	Maxillary bone free posteriorly, not covered by preorbital; body compressed laterally	*Go to* **4**
	Maxillary bone covered posteriorly by preorbital; body fusiform	*Go to* **7**
4	Base of 1st dorsal fin longer than base of 2nd; ventral fins set close together, their interspace not exceeding 66% of ventral fin width at base; teeth small, none of them canines	*Perca fluviatilis* p226
	Base of 1st dorsal fin equal to, or less than, base of 2nd; space between ventral fins at least 66% of ventral fin width at base	*Go to* **5**
5	Never more than 18 branched rays in dorsal fins; interobital width much greater than eye diameter; canines present	*Sander marina* p228
	More than 18 branched rays in dorsal fins; interorbital width shorter than, or equal to, eye diameter	*Go to* **6**
6	Canines well developed; cheeks naked or only partly scaled; 80–97 scales along lateral line	*Sander lucioperca* p228
	No canines in adults; cheeks completely scaled; 70–83 scales along lateral line	*Sander volgensis* p229
7	1st dorsal fin with 13–15 spines; 2nd dorsal fin with 18–20 rays	*Zingel zingel* p230
	1st dorsal fin with 8–9 spines; 2nd dorsal fin with 12–13 rays	*Go to* **8**
8	Caudal peduncle as long as, or longer than, base of 2nd dorsal fin; 4–5 distinct dark bars across body	*Zingel streber* p230
	Caudal peduncle shorter than base of 2nd dorsal fin; 3 irregular dark bars across body	*Zingel asper* p229
9	Marked space between dorsal fins; maxillary free posteriorly	*Percarina demidoffi* p227
	Dorsal fins united; maxillary covered by preorbital	*Go to* **10**
10	Snout short, the same length as, or shorter than, eye diameter; dorsal fin with 11–16 spines; 35–40 lateral scales	*Go to* **11**
	Snout elongate, at least half as long again as eye diameter; dorsal fin with 17–19 spines; 50–62 scales along lateral line	*Go to* **12**
11	dorsal fin with 11–16 spines; 35–40 scales along lateral line; no marked dark lateral bars	*Gymnocephalus cernuus* p225
	Dorsal fin with 14–16 spines; 35–39 scales along lateral line; several marked dark lateral bars	*Gymnocephalus baloni* p224
12	Body length less than 5 times greatest depth; round black spots present on sides of body	*Gymnocephalus acerina* p224
	Body length more than 5 times greatest depth; 3 or 4 black transverse bands on sides of body	*Gymnocephalus schraetser* p225

Don Ruffe *Gymnocephalus acerina*

Size 12–18 cm; maximum 21 cm. **Distinctive features** two dorsal fins partially joined to one another, the anterior with 17–19 spines; 50–55 lateral scales; small dark spots on sides but no stripes. **Distribution** found only in slow-flowing waters and lakes in river basins entering the north of the Black Sea (e.g. Dniester, Dnieper, Don). **Reproduction** April–May, spawning among stones and vegetation in shallow water. **Food** invertebrates, especially crustaceans and insects; sometimes small fish. **Value** of little commercial or sporting value. **Conservation Status** Lower Risk.

Balon's Ruffe *Gymnocephalus baloni*

Size 15–20 cm; maximum 25 cm. **Distinctive features** 2 small spines on operculum; several dark lateral vertical bars; 14–16 anterior dorsal rays; 35–39 lateral scales. **Distribution** found only in slow-flowing parts of the River Danube and associated tributaries. **Reproduction** April–May, spawning in shallow water among stones and vegetation. **Food** bottom-dwelling invertebrates, fish eggs and (when larger) small fish. **Value** of little commercial or sporting value. **Conservation Status** Vulnerable.

Ruffe *Gymnocephalus cernuus*

Size 10–15 cm; maximum 30 cm (exceptionally 50 cm in Siberia); British rod record 113 g. **Distinctive features** two dorsal fins partially joined to one another, the anterior with 11–16 spines; 35–40 lateral scales. **Distribution** found in slow-flowing rivers and lakes throughout much of Europe, except certain areas in the north and south. Introduced to, and expanding rapidly in, Lake Superior through the discharge of ballast water from ships entering the Great Lakes from Europe. **Reproduction** April–May, spawning among stones and vegetation in shallow water. The eggs hatch in 8–12 days and the young mature after 2–3 years. 4,000–104,000 eggs per female. **Food** bottom-dwelling invertebrates (especially molluscs, crustaceans and insect larvae) and sometimes small fish. **Value** of minor commercial and sporting value, although it is caught in a few areas and used as live bait in others. **Conservation Status** Lower Risk.

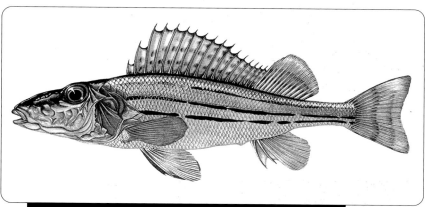

Striped Ruffe *Gymnocephalus schraetser*

Size 15–20 cm; maximum 24 cm. **Distinctive features** two dorsal fins partially joined to one another, the anterior with 17–19 spines; 55–62 lateral scales; several dark longitudinal stripes on sides. **Distribution** found only in the Danube basin, and occasionally in its estuary to the Black Sea. **Reproduction** April–May, spawning in shallow water among stones and vegetation. **Food** invertebrates (mainly molluscs, crustaceans and insects) but some fish when larger. **Value** of little commercial or sporting value. **Conservation Status** Endangered.

European Perch *Perca fluviatilis*

Size 20–35 cm; maximum 51 cm; maximum weight about 4.75 kg; Scottish rod record 2.211 kg (1989). **Distinctive features** separate dorsal fins, the first very spiny; reddish pectoral and pelvic fins; several dark vertical stripes on sides. **Distribution** in slow rivers and lakes over most of northern Europe and Asia. The Yellow Perch *Perca flavescens* of eastern North America is believed by some to be the same species. **Reproduction** April–June, in shallow water among vegetation. The whitish eggs laid in ribbons about 1 m long hatch in 15–20 days. Young mature in 2–3 years and live to 10 years. 12,000–199,000 eggs per female. **Food** invertebrates (especially crustaceans and insect larvae) when young, invertebrates and fish when older. **Value** caught commercially in some countries by traps, nets and baited lines. Also very important as a sport species. **Conservation Status** Lower Risk.

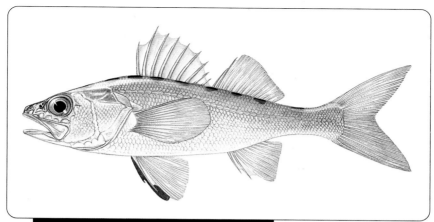

Percarina *Percarina demidoffi*

Size 6–9 cm; maximum 10 cm. **Distinctive features** 2 well-separated dorsal fins, the anterior with 9–11 spines; a row of black spots along lateral line which has 33–37 scales. **Distribution** found only in the northern area of the Black Sea, in the Sea of Azov, and in the lower reaches of associated rivers. **Reproduction** June–July, spawning in shallow areas in both fresh and sea water over sand and silt. The eggs hatch in 2 days and the young mature after 1 year. The adults rarely live longer than 3–4 years. A maximum of about 3,000 eggs per female. **Food** invertebrates, especially crustaceans and small fish. **Value** of some commercial importance in local net fisheries. **Conservation Status** Vulnerable.

Asprete *Romanichthys valsanicola*

Size 8–12 cm; maximum 13 cm. **Distinctive features** 2 well-separated dorsal fins, the posterior larger than the anterior; eyes positioned dorsally on head; 58–67 lateral scales. **Distribution** discovered as recently as the 1950s, and found only in fast-flowing water in the upper reaches of certain rivers (Arges, Vilsan and Riul) in the Danube basin in Rumania. **Reproduction** reproductive behaviour unknown. **Food** invertebrates (mainly insect larvae, especially stoneflies) and small fish. **Value** of no commercial or sporting value. **Conservation Status** Critically Endangered.

Pikeperch *Sander lucioperca*

Also called Zander. **Size** 30–70 cm; maximum 130 cm; maximum weight about 18 kg; British rod record 6.945 kg. **Distinctive features** dorsal fins almost touching; large canine teeth; 80–95 lateral scales. **Distribution** slow rivers and rich lakes in Europe from the Netherlands to the Caspian Sea. Introduced to other areas (e.g. England). **Reproduction** April–June, among gravel and stones. The eggs are guarded by both parents and hatch in 5–10 days. Young mature in 3–5 years. 180,000–1,185,000 eggs per female. **Food** invertebrates at first, but almost entirely fish thereafter. **Value** taken commercially in traps and nets and a valued food fish. A prized sport species. **Conservation Status** Lower Risk.

Sea Pikeperch *Sander marina*

Size 30–50 cm; maximum 62 cm; specimens 54 cm long weigh about 1.7 kg. **Distinctive features** dorsal fins almost touching; large canine teeth; 78–84 lateral scales. **Distribution** Black and Caspian Seas and the lower reaches of rivers there (e.g. Bug and Dnieper). **Reproduction** April–May, over stones in fresh and salt water. Young mature after 3–5 years and live to 10 years. 13,000–126,000 eggs per female. **Food** invertebrates initially, but almost entirely fish afterwards. **Value** of commercial significance in trap and net fisheries. Rarely angled for. **Conservation Status** Vulnerable.

Volga Pikeperch *Sander volgensis*

Size 25–40 cm; maximum 45 cm; mean weight about 1.3 kg. **Distinctive features** dorsal fins almost touching; no canine teeth in adults; 70–83 lateral scales. **Distribution** in the basins of rivers entering the Black and Caspian Seas (e.g. Volga and Danube): Occurs in slow rivers and lakes. **Reproduction** April–May, among stones and vegetation. Young mature after 3–4 years. **Food** large invertebrates, especially crustaceans, and fish of various kinds. **Value** caught commercially in nets and traps. Of little angling importance. **Conservation Status** Vulnerable.

Asper *Zingel asper*

Size 15–20 cm; maximum 22 cm. **Distinctive features** 2 well-separated dorsal fins, the anterior with 8–9 spines, the posterior with 12–13 rays; caudal peduncle shorter than base of posterior dorsal fin. **Distribution** found only in the River Rhône and its tributaries. **Reproduction** March–April, spawning among stones and vegetation in shallow water. **Food** mainly bottom-dwelling invertebrates (crustaceans and insects especially), but also fish eggs and fry. **Value** of no commercial or sporting value. **Conservation Status** Endangered.

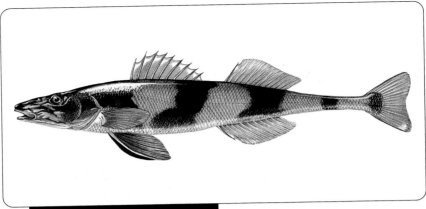

Streber *Zingel streber*

Size 12–15 cm; maximum 22 cm. **Distinctive features** 2 well-separated dorsal fins, the anterior with 8–9 spines, the posterior with 12–13 rays; caudal peduncle as long as, or longer than, base of posterior dorsal fin. **Distribution** found only in running water in the River Danube and its tributaries (usually at higher altitudes than the Zingel), and in the Vardar River which flows into the Aegean Sea. **Reproduction** March–April, among stones in fast-flowing water. **Food** invertebrates (mainly crustaceans and insect larvae) and some small fish. **Value** of little commercial or sporting value. **Conservation Status** Endangered.

Zingel *Zingel zingel*

Size 15–30 cm; maximum 48 cm. **Distinctive features** 2 well-separated dorsal fins, the anterior with 13–15 spines, the posterior with 18–20 rays; caudal peduncle shorter than base of posterior dorsal fin. **Distribution** only in running water in the basin of the River Danube (usually at lower altitudes than the Streber). **Reproduction** March–May, in fast-flowing water among stones. About 5,000 eggs per female. **Food** benthic invertebrates (especially crustaceans and insect larvae) and small fish. **Value** of no commercial or sporting value. **Conservation Status** Vulnerable.

Cichlid Family
Cichlidae

The Cichlidae is a large family of fish, many of them extremely colourful, which is native mainly to tropical Africa, South America and Central America – not to Europe. They occur in a variety of habitats from large overgrown rivers to enormous deep lakes. They have only 1 pair of nostrils and the lateral line is usually in two parts. Breeding normally involves elaborate sexual display, nest building and parental care. Although common in tropical aquaria, only 1 species, the Chanchito, has been introduced successfully to natural waters in Europe – though there are unsubstantiated reports of populations of other species (e.g. Oscar *Astronotus ocellatus*, Tilapia *Tilapia zillii*).

Chanchito *Cichlasoma facetum*

Size 10–15 cm; maximum 30 cm. **Distinctive features** 1 pair of nostrils; large dorsal fin with erect spiny anterior rays; colour variable – brassy-yellow to greenish or even black, usually several dark transverse bands. **Distribution** native to South America, and known to tolerate much lower temperatures than most of the family, it has been successfully introduced to southern Portugal. **Reproduction** June–August, spawning in a nest previously prepared by both male and female. Eggs hatch in 2–4 days and are protected by the parents for some time. **Food** invertebrates, especially molluscs and insect larvae. **Value** of no sporting or real commercial value other than its significance as an aquarium fish. **Conservation Status** Alien.

Blenny Family
Blenniidae

The **Blenniidae, or blennies,** are mainly small marine fish, bottom living in habit. There are about 280 species, 20 of which occur in Europe. They are usually elongate, slender, scaleless fish with long dorsal fins stretching from head to tail, and slender pelvic fins anterior to the broad pectorals. The bodies are characteristically marked with coloured blotches or stripes. Some have elaborate tentacles projecting above the eye.

A number of species occur around European coasts, mainly in the Mediterranean, but only 1 is found in fresh water.

Freshwater Blenny *Salaria fluviatilis*

Size 8–12 cm; maximum 15 cm. **Distinctive features** an enormous dorsal fin running from head to tail; small tentacle just above each eye; pelvic fins anterior to pectoral fins. **Distribution** found on stony and muddy bottoms in fresh water in streams and lakes of basins associated with the western and north-eastern Mediterranean. **Reproduction** April–June, spawning in nests under stones. Eggs are guarded by the male. **Food** invertebrates (especially crustaceans and insects) and small fish. **Value** of no commercial and little angling importance, except as a bait fish. **Conservation Status** Endangered.

Goby Family
Gobiidae

The Gobiidae, or gobies, is a very successful family of small fish found in many parts of the world, both tropical and temperate. Most species are marine but many occur in brackish water and there is a considerable number of freshwater species. These occur in a variety of habitats, in both running and standing waters. Because of close relationships within genera, and the small size of many species, identification may often be difficult; it is further complicated by the fact that there are often considerable differences between the sexes.

The main characteristic of this family is the fact that the pelvic fins are united to form a single round or oval sucker-like fin with a transverse anterior membrane. The body is normally rather elongate, although often broad and squat anteriorly. The head is large and the lips and cheeks well developed. Sensory papillae and other protuberances are common on the head and often arranged in characteristic patterns. The lateral line on the body is either incomplete or absent. There are two well-developed dorsal fins, but neither has strong spiny rays. The anal fin is usually similar in size to the second dorsal fin. Characteristics important in identification include the extent of sensory (lateral-line) canals on the head, as well as patterns of sensory papillae there. Meristic characters (e.g. numbers of vertebrae) and adult coloration are often important.

Gobies are often very abundant fish, usually benthic in habit and common in shallow coastal areas. A number of species are pelagic. At spawning time, the males normally become much darker (some turn pure black), their fins elongate and the shape of the head alters. They build simple nests – usually in the shelter of shells, stones or weed, and guard the adhesive eggs (which are usually pear-shaped with attachment filaments at one end), laid in patches until the larvae emerge and swim away.

This is a large and important family in both the sea and fresh water with some uncertainty remaining over the status of the various species and subspecies. Altogether in Europe, some 30 species occur in fresh water or enter it from time to time in their life history but only 26 of these are discussed fully below. Some work remains to be done on the taxonomy of those which are not dealt with in full here (i.e. *Knipowitschia goerneri, Knipowitschia milleri, Knipowitschia mrakovcici, Knipowitschia thessala*) and there may still be a few species to be discovered; those covered here in full may be identified as follows:

Key to Gobiidae in Europe

	Perianal organ present	*Go to* **1**
	Perianal organ absent	*Go to* **2**
1	Perianal organ large, more than 50% of pelvic disc; pectoral fins with 17–18 rays; adult length more than 3 cm	***Economidichthys pygmaeus*** p237
	Perianal organ small, less than 50% of pelvic disc; pectoral fins with 13–16 rays; adult length less than 3 cm	***Economidichthys trichonis*** p238
2	Body generally without scales; although there may be platelets or spikes; no sensory canals or pores on head	*Go to* **3**
	Body covered with scales; sensory canals and pores present on head	*Go to* **7**
3	Body entirely naked; anterior nostrils not produced into tubules	***Caspiosoma caspium*** p237
	Body normally covered with platelets or spikes; anterior nostrils produced into tubules	*Go to* **4**
4	6 rays in 1st dorsal fin, 12–13 rays in 2nd dorsal fin, 9–11 rays in anal fin; sparse spiny scales on side of body	***Benthophiloides brauneri*** p235
	1–4 rays in 1st dorsal fin, 6–11 rays in 2nd dorsal fin, 6–10 rays in anal fin; no scales on side of body	*Go to* **5**

5   Body covered in uniform body granules, but without rows of larger granules   *Benthophilus granulosus* p235

3 rows of large spines on each side of body, usually with small bony granules between; abdomen and pectoral regions naked   *Go to 6*

---

6   Body densely covered with bony granules; 22–27 tubercles in the dorsal series; colour ash-grey, with no dark bands across back   *Benthophilus macrocephalus* p236

Body covered mainly with tubercles; very few bony granules present; 3 brownish bands across back.   *Benthophilus stellatus* p236

---

7   2nd dorsal fin with less than 12 branched rays   *Go to 8*

2nd dorsal fin with more than 12 branched rays   *Go to 16*

---

8   Anterior nostrils produced into barbel-shaped tubercles overhanging mouth; 36–48 lateral scales.   *Proterorhinus marmoratus* p246

Anterior nostrils not produced into barbel-shaped tubules; 45–79 lateral scales   *Go to 9*

---

9   Swim-bladder present   *Zosterisessor ophiocephalus* p247

No swim-bladder present in adults   *Go to 10*

---

10   Parietal and occipital areas not scaled   *Go to 11*

Parietal and, usually, occipital areas scaled   *Go to 15*

---

11   Anal fin with 10–13 branched rays; 42–58 lateral scales   *Neogobius melanostomus* p243

Anal fin with 11–17 branched rays; more than 55 lateral scales   *Go to 12*

---

12   2nd dorsal fin becoming lower posteriorly   *Neogobius fluviatilis* p242

2nd dorsal fin never becoming lower posteriorly, being either higher at the ends or of equal height throughout   *Go to 13*

---

13   Caudal peduncle about 1 1/2 times as long as deep; lateral lobules on collar of ventral sucker obtuse   *Neogobius syrman* p244

Caudal peduncle as long as deep, or only slightly longer; lateral lobules on collar of ventral sucker pointed   *Go to 14*

---

14   Minimum body depth more than 8 per cent of body length; thickness of caudal peduncle less than 66 % of its depth   *Neogobius cephalarges* p241

Minimum body depth less than 8 per cent of body length; thickness of caudal peduncle more than 66 % of its depth   *Neogobius kessleri* p243

---

15   65–84 lateral scales; 8 or more pits in suborbital series   *Mesogobius batrachocephalus* p241

47–69 lateral scales; 6 or more pits in suborbital series   *Neogobius gymnotrachelus* p242

---

16   More than 60 lateral scales   *Relictogobius kryzanovskii* p247

Less than 60 lateral scales   *Go to 17*

---

17   Anal fin with 7 branched rays; 35–40 scales along lateral line   *Go to 18*

Anal fin with 8–11 branched rays   *Go to 20*

---

18   Less than 29 lateral scales; 8 branched rays in 2nd dorsal fin   *Knipowitschia panizzae* p240

29–50 lateral scales; 9 branched rays in 2nd dorsal fin   *Go to 19*

---

19   29–44 lateral scales; no head lateral line canal   *Padogobius martensii* p244

40–50 lateral scales; head lateral line canal present   *Padogobius nigricans* p245

---

20   30 lateral scales; 8–9 branched rays in anal fin   *Hyrcanogobius bergi* p238

More than 30 lateral scales; 8–11 branched rays in anal fin   *Go to 21*

---

21   Space between dorsal fins about equal to length of 1st dorsal fin   *Knipowitschia longecaudata* p239

Space between dorsal fins much less than length of 1st dorsal fin   *Go to 22*

---

22   More than 40 lateral scales   *Pomatoschistus microps* p246

Less than 40 lateral scales   *Go to 23*

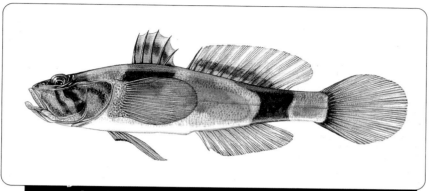

Banded Tadpole Goby *Benthophiloides brauneri*

Size 4–5 cm; maximum 7 cm. **Distinctive features** 2 dorsal fins and united pelvic fins; no scales when adult; 2 dark brownish bands around body; 2 oblique streaks on cheeks. **Distribution** found in the Caspian Sea and the lower reaches of the Dnieper and Bug. **Reproduction** spawning takes place in shallow brackish water, but little is known of its reproductive habits. **Food** invertebrates, especially crustaceans. **Value** of no commercial or sporting importance. **Conservation Status** Lower Risk.

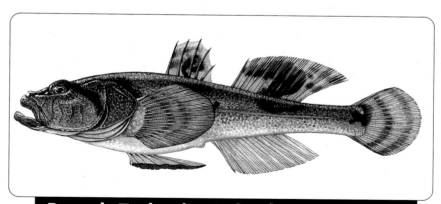

Rough Tadpole Goby *Benthophilus granulosus*

Size 3–4 cm; maximum 6 cm. **Distinctive features** 2 dorsal fins and united pelvic fins; second dorsal fin with 6–9 and anal fin with 6–8 branched rays; head and body covered in uniform bony granules, but without rows of larger granules. **Distribution** common throughout the Caspian Sea and in the delta and estuary of the Volga. **Reproduction** nothing is known of the reproductive habits of this species. **Food** invertebrates, especially crustaceans, and some small fish larvae. **Value** of no commercial or sporting value. **Conservation Status** Lower Risk.

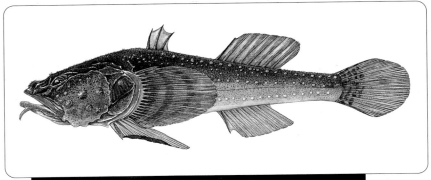

Caspian Tadpole Goby
Benthophilus macrocephalus

Size 5–10 cm; maximum 12 cm. **Distinctive features** 2 dorsal fins and united pelvic fins; second dorsal fin with 7–9 branched rays; head and body densely covered in bony granules; 23–25 tubercles in the dorsal series. **Distribution** found throughout the Caspian Sea and in the Sea of Azov. Common in lagoons and the estuaries of the Volga and the Kuban. **Reproduction** little is known of the reproductive habits of this species. **Food** invertebrates, mainly worms, molluscs, crustaceans and insect larvae. **Value** of no commercial or sporting significance. **Conservation Status** Lower Risk.

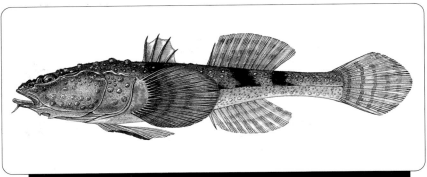

Stellate Tadpole Goby Benthophilus stellatus

Size 8–12 cm; maximum 14 cm. **Distinctive features** head broad and flat; 2 dorsal fins and united pelvic fins; second dorsal fin with 7–8 branched rays; 28 vertebrae; large numbers of tubercles (but no scales) in characteristic patterns on head and body; 3 dark spots on back. **Distribution** found in brackish water areas of the Black Sea, the Sea of Azov and the Caspian Sea. Commonly found upstream in rivers in the Black Sea area (Danube, Dnieper, Don), but does not enter fresh water in the Caspian basin. **Reproduction** May–June, after which all the adult males and females die; thus the whole life span is exactly 1 year. 700–2,500 eggs per female. **Food** invertebrates, mainly worms and molluscs and small fish. **Value** of no commercial or sporting significance. **Conservation Status** Lower Risk.

Caspian Goby *Caspiosoma caspium*

Size 3–4 cm; maximum 5 cm. **Distinctive features** 2 dorsal fins and united pelvic fins; body entirely naked; anterior nostrils produced into tubules; second dorsal fin with 11–13 and anal fin with 8–10 branched rays. **Distribution** found in the northerly areas of the Black Sea (including the Sea of Azov) and the Caspian Sea. Common in the lower reaches of large rivers (e.g. Volga, Don, Dnieper) in these areas. **Reproduction** large eggs, but little is known of the reproductive habits of this species. **Food** invertebrates, mainly crustaceans, and (in fresh water) insect larvae. **Value** of no commercial or sporting value. **Conservation Status** Lower Risk.

Economidis' Goby *Economidichthys pygmaeus*

Size 3–4 cm; maximum 4.5 cm. **Distinctive features** body dark with transverse lateral bars; perianal organ more than 50% of pelvic disc; pectoral fins with 17–18 rays. **Distribution** found only in western Greece on the islands of Lefkas, Epiris and Lake Trichonis, in both running and still water with abundant vegetation over silt. **Reproduction** April–July, with more than one spawning per season. **Food** not known, but probably small invertebrates. **Value** of no commercial or angling value. **Conservation Status** Vulnerable.

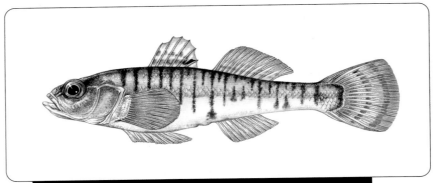

Trichonis Goby *Economidichthys trichonis*

Size 1.5–2.5 cm; maximum 3 cm. This may be the smallest freshwater European teleost at maturity (female 1.8 cm) and maximum size (male 3.0 cm). **Distinctive features** body pale with vertical stripes; perianal organ less than 50% of pelvic disc; pectoral fins with 13–16 rays. **Distribution** found only in Greece in Lake Trichonis, mainly at the mouths of inflowing streams. **Reproduction** April–July, with more than one spawning per season. **Food** not known, but probably small invertebrates. **Value** of no commercial or angling value. **Conservation Status** Lower Risk.

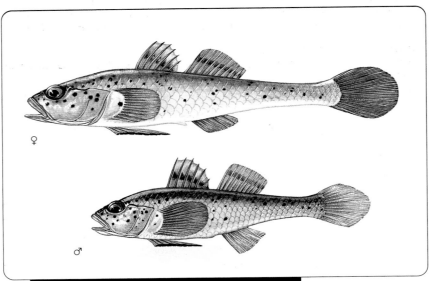

Berg's Goby *Hyrcanogobius bergi*

Size 2–3 cm; maximum 3.5 cm; this is one of the smallest European fish. **Distinctive features** 2 dorsal fins and united pelvic fins; more than 30 lateral scales; dorsal fin with 8–9 and anal fin with 8–9 branched rays. **Distribution** found only in the northern Caspian Sea, especially in or near the estuaries of the Volga, Emba and Ural Rivers. **Reproduction** July–September, laying eggs under stones and shells. Mature at 2 years. **Food** invertebrates, especially crustaceans. **Value** of no commercial or sporting value. **Conservation Status** Lower Risk.

Caucasian Goby *Knipowitschia caucasica*

Size 2–3 cm; maximum 4 cm. **Distinctive features** 2 dorsal fins and united pelvic fins; 31–38 lateral scales; second dorsal fin with 8 branched rays; 30–32 vertebrae; no longitudinal stripes on dorsal fins. **Distribution** found all round the coasts of the Black Sea (including the Sea of Azov) and the Caspian Sea, and in estuaries, lagoons and easily accessible fresh waters. **Reproduction** March–July, eggs laid under stones or shells; 220–500 eggs per female. **Food** invertebrates, especially crustaceans, and (in fresh water) insect larvae. **Value** of no commercial or sporting value. **Conservation Status** Lower Risk.

Korission Goby *Knipowitschia goerneri*
recorded from Greece (type locality: springs near Korissias lagoon). **Conservation Status** Endangered.

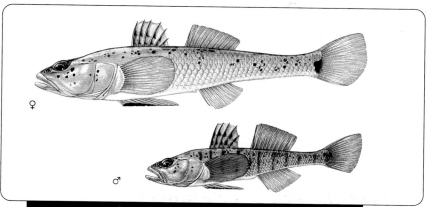

Longtail Goby *Knipowitschia longecaudata*

Size 3–4 cm; maximum 5 cm. **Distinctive features** 2 dorsal fins and united pelvic fins; 36–43 lateral scales; 32–34 vertebrae; space between dorsal fins about equal to length of 1st dorsal fin. **Distribution** found in brackish waters in the Black Sea (including the Sea of Azov) and the Caspian Sea, and in lower reaches of several large rivers (e.g. Dnieper, Don). **Reproduction** nothing appears to be known about the reproductive habits of this species. **Food** invertebrates, especially crustaceans. **Value** of no commercial or sporting value. **Conservation Status** Lower Risk.

Goby Family Gobiidae

Miller's Goby *Knipowitschia milleri*
recorded from Greece (type locality: River Acheron delta). **Conservation Status** Lower Risk.

Mracovcic's Goby *Knipowitschia mrakovcici*
recorded recently from Croatia (Krka River basin). **Conservation Status** Endangered.

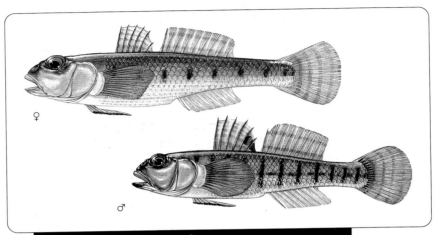

Panizzae's Goby *Knipowitschia panizzai*

Size 2–3 cm; maximum 5 cm. **Distinctive features** 2 dorsal fins (the second with 8 branched rays), and united pelvic fins; 32–39 lateral scales; 29–30 vertebrae. **Distribution** found only in northern Italy in brackish estuaries and lagoons; also rivers near Venice (e.g. River Po) and in Lake Garda and Lake Maggiore. **Reproduction** March–August, eggs laid under stones and shells; mature after 1 year. **Food** invertebrates, particularly worms, crustaceans and insect larvae. **Value** of no commercial or sporting value. **Conservation Status** Vulnerable.

Orsini's Goby *Knipowitschia punctatissima*

Size 4–5 cm; maximum 6 cm. **Distinctive features** body naked except for a few scales behind the pectoral fins; 31 vertebrae; males with several dark bars, female with chin spot. **Distribution** found in springs and upland streams in northern Italy and Croatia. **Reproduction** April–June, in waters where the current is slow and water temperature constant. 300 eggs per female. Lives for 2–3 years. **Food** not known, but probably benthic invertebrates. **Value** of no commercial or sporting value. **Conservation Status** Vulnerable.

Thessaly Goby *Knipowitschia thessala*
recorded from Greece (type locality: Kefalovriso Spring). **Conservation status** Vulnerable.

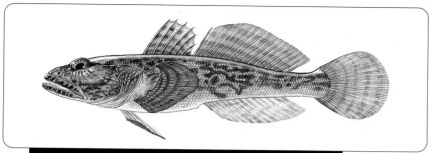

Toad Goby *Mesogobius batrachocephalus*

Size 25–30 cm; maximum 35 cm – the largest European goby. **Distinctive features** 2 dorsal fins and united pelvic fins; 65–84 lateral scales; second dorsal fin with 16–19 branched rays; 34–36 vertebrae; 8 or more pits in suborbital series. **Distribution** found along the northern coasts of the Black Sea, including the Sea of Azov. Occurs in brackish lagoons in estuarine areas but only occasionally enters rivers (e.g. Bug, Dnieper). **Reproduction** February–May, in coastal waters. The large eggs (5 mm) are laid in a nest among stones, prepared and guarded by the male. **Food** invertebrates, particularly crustaceans, and some small fish. **Value** a significant commercial species in net fisheries in the Black Sea. **Conservation Status** Lower Risk.

Ginger Goby *Neogobius cephalarges*

Size 18–22 cm; maximum 24 cm. **Distinctive features** 2 dorsal fins and united pelvic fins; second dorsal fin with 15–21 branched rays; 48–79 lateral scales; pointed lateral lobules on collar of ventral sucker. **Distribution** found in shallow stony areas in the Black and Caspian Seas (including the Sea of Azov) and in many rivers associated with them. Many of the fresh water populations are non-migratory and spend all their lives in clear stony mountain streams. **Reproduction** March–May in the sea. Little is known about its reproductive habits. **Food** invertebrates (mainly molluscs and crustaceans) and small fish. **Value** of some commercial value to local sea fishing. Not angled for. **Conservation Status** Lower Risk.

Monkey Goby *Neogobius fluviatilis*

Size 15–18 cm; maximum 20 cm. **Distinctive features** 2 dorsal fins (the second fin becoming lower posteriorly), and united pelvic fins; 55–61 lateral scales; second dorsal fin with 14–17 branched rays; 32–34 vertebrae. **Distribution** found in some brackish areas of the Black Sea (including the Sea of Azov) and the Caspian Sea, but most common in larger rivers in these areas (e.g. Danube, Dniester, Bug). **Reproduction** April–September, spawning in a nest prepared and guarded by the male. The young mature after 2 years; lives to 4 years. **Food** invertebrates, especially worms and crustaceans and, in fresh water, insect larvae. **Value** of considerable commercial value in both the Black and the Caspian Sea areas, where some of the catch is canned. **Conservation Status** Vulnerable.

Racer Goby *Neogobius gymnotrachelus*

Size 12–15 cm; maximum 17 cm. **Distinctive features** 2 dorsal fins and united pelvic fins; second dorsal fin with 14–18 branched rays; 54–69 lateral scales; 6 or more pits in suborbital series. **Distribution** found in several streams and a few lakes (e.g. Lake Paleostom) in river basins north of the Black Sea (e.g. Danube [up to Bucharest], Dniester, Bug). It also occurs in brackish water here and throughout the Caspian Sea. **Reproduction** April–May, spawning in a nest prepared and subsequently guarded by the male. **Food** various invertebrates, especially crustaceans. **Value** of minor value to commercial net fishermen in the Black Sea. Of no angling importance. **Conservation Status** Lower Risk.

Bighead Goby *Neogobius kessleri*

Size 15–20 cm; maximum 22 cm. **Distinctive features** 2 dorsal fins and united pelvic fins; second dorsal fin with 15–19 branched rays; 64–79 lateral scales; pointed lateral lobules on collar of ventral sucker. **Distribution** found in the Black and Caspian Seas and in the basins of their northern rivers. Occurs commonly in the sea, in brackish water, in fast-flowing water and in lakes (e.g. Lakes Brates and Kagul). **Reproduction** April–May, spawning in a nest which is subsequently guarded by the male; 1,500–2,000 eggs per female. **Food** invertebrates, especially crustaceans. **Value** of some commercial importance to net fishermen in the Black and Caspian Seas. Of no angling importance. **Conservation Status** Lower Risk.

Round Goby *Neogobius melanostomus*

Size 18–22 cm; maximum 25 cm. **Distinctive features** 2 dorsal fins and united pelvic fins; anal fin with 10–13 branched rays; 45–57 lateral scales. **Distribution** found in the Black and Caspian Seas, and in the lower reaches of large river systems associated with these (e.g. Dniester, Volga). Introduced to Baltic Sea and also to North America where it is established in the Great Lakes. **Reproduction** mainly May–July, but may extend well outside this in different areas. Spawns in both fresh and salt water. The young mature after 1 year. **Food** invertebrates, especially molluscs and crustaceans. **Value** an abundant species and of considerable local commercial value in the Black Sea and the Caspian Sea where the catch may be marketed either fresh or canned. **Conservation Status** Lower Risk.

Syrman Goby *Neogobius syrman*

Size 18–22 cm; maximum 25 cm. **Distinctive features** 2 dorsal fins and united pelvic fins; second dorsal fin with 15–19 branched rays; 57–78 lateral scales; lateral lobules on collar of ventral sucker obtuse. **Distribution** found in the northern areas of the Black Sea (including the Sea of Azov) and the Caspian Sea. Enters estuaries and the lower reaches of rivers in these areas (e.g. Bug, Don) and some lakes (e.g. Lake Razelm). **Reproduction** April–May, spawning in a nest which is subsequently guarded by the male. **Food** invertebrates, especially crustaceans. **Value** of some commercial importance to net fishermen in the Black and Caspian Seas. Of no angling value. **Conservation Status** Lower Risk.

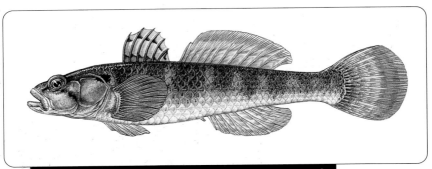

Martens' Goby *Padogobius martensii*

Size 7–8 cm; maximum 9 cm. **Distinctive features** nape and head naked; no anterior nostril process; 29–44 lateral scales; 28–29 vertebrae. **Distribution** occurs in rivers and small streams and at the edges of lakes in Italy, Croatia and Slovenia. **Reproduction** May–July, spawning in a nest under a stone or in a hole. Short lived. **Food** benthic invertebrates, especially crustaceans and insects. **Value** of no commercial or angling value. **Conservation Status** Lower Risk.

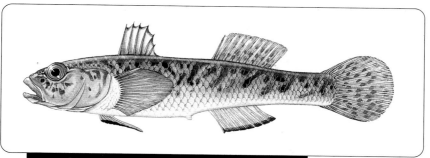

Italian Goby *Padogobius nigricans*

Size 5–10 cm; maximum 13 cm. **Distinctive features** 2 dorsal fins (the second with 11–17 branched rays), and united pelvic fins; 40–50 lateral scales; 29–30 vertebrae. **Distribution** found only in western Italy in clear running waters in the basins of the rivers Tiber and Arno. **Reproduction** May–June, spawning under stones in shallow running water. **Food** invertebrates, especially insect larvae. **Value** of no commercial or sporting value. **Conservation Status** Vulnerable.

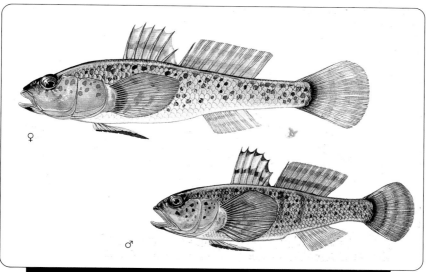

Canestrinii's Goby *Pomatoschistus canestrini*

Size 4–5 cm; maximum 7 cm. **Distinctive features** 2 dorsal fins and united pelvic fins; 36–42 lateral scales; second dorsal fin with 8–9 branched rays; 30 vertebrae; longitudinal stripes on both dorsal fins. **Distribution** found in brackish waters, lagoons and fresh waters in the former Yugoslavia (River Jodro) and Italy. **Reproduction** March–July, mature at 1 year. **Food** invertebrates, especially worms, crustaceans and insect larvae. **Value** of no commercial or sporting value. **Conservation Status** Vulnerable.

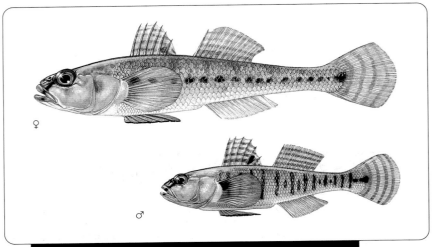

♀

♂

Common Goby *Pomatoschistus microps*

Size 3–6 cm; maximum 7 cm. **Distinctive features** 2 dorsal fins and united pelvic fins; 39–52 lateral scales; second dorsal fin with 8–11 branched rays; 30–32 vertebrae; space between dorsal fins much less than length of 1st dorsal fin. **Distribution** found in shallow, often brackish water lagoons round the coasts of Europe from southern Norway to the Mediterranean and Black Seas. Common in estuaries. **Reproduction** April–August, spawning in a nest cleared by the male under a shell or stone. The male guards the eggs and fans water over them until they hatch. Fish may spawn up to 8 times in a breeding season; 3,400 eggs per female. The young mature after 1 year, and few fish live beyond 2 years. **Food** mainly benthic invertebrates, especially worms and crustaceans. **Value** of no commercial or sporting value. **Conservation Status** Lower Risk.

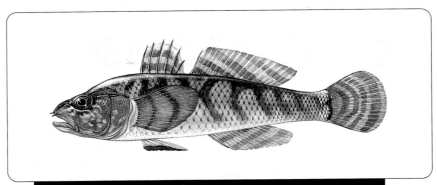

Tubenose Goby *Proterhorinus marmoratus*

Size 5–10 cm; maximum 11 cm. **Distinctive features** 2 dorsal fins (the second with less than 12 branched rays), and united pelvic fins; 36–48 lateral scales. **Distribution** found in shallow water round most of the coasts of the Black and Caspian Seas. Common in lagoons, some lakes and the lower reaches of many rivers flowing into these (e.g. Danube, Bug, Araks). Accidentally introduced via ballast water to North America, it is now established in the Great Lakes. **Reproduction** eggs laid under stones or shells and guarded by male. **Food** invertebrates, mainly worms and crustaceans, and (in fresh water) insect larvae. **Value** of no commercial or sporting value. **Conservation Status** Lower Risk.

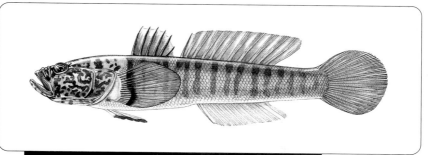

Relict Goby *Relictogobius kryzanovskii*

Size 4–6 cm; maximum 7 cm. **Distinctive features** 2 dorsal fins (the second with 10–11 branched rays); united pelvic fins; tail rounded; 63–72 lateral scales. **Distribution** found only in a salt lake on the coast of the Black Sea in the region of Novorossiisk. **Reproduction** little is known of the reproductive habits of this species. **Food** invertebrates, mainly crustaceans. **Value** of no commercial or sporting value. **Conservation Status** Lower Risk.

Grass Goby *Zosterisessor ophiocephalus*

Size 18–22 cm; maximum 25 cm. **Distinctive features** 2 dorsal fins and united pelvic fins; 53–68 lateral scales; second dorsal fin with 13–16 branched rays; swim-bladder present. **Distribution** found in brackish lagoons in the northern Mediterranean and the western and eastern Black Sea (including the Sea of Azov). Only occasionally occurs in fresh water. **Reproduction** February–July, spawning in nests built by the male from plant detritus, among thick vegetation; 45,000 eggs per female; matures at 2–3 years. **Food** invertebrates, especially crustaceans. Sometimes small fish. **Value** of some commercial value to net fisheries in the seas concerned. **Conservation Status** Lower Risk.

Sleeper Family
Eleotridae

The Eleotridae, or sleepers, are an interesting family of fish related to the gobies and there are many genera. They occur in fresh and brackish water in several tropical and temperate regions of the world, though none are native to Europe.

Only one, introduced, species is found in fresh water in Europe.

Chinese Sleeper *Perccottus glehni*
Size 7–9 cm; maximum 25 cm. **Distinctive features** 37–43 lateral scales; anal fin with 7–10, ventral with 5 branched rays. Dark brown spots on sides and speckles on abdomen. **Distribution** eastern Russia and North Korea. Introduced to Europe and recently established in catchments in western Russia, Poland (Vistula) and Hungary (Tisza). Usually found in lakes, ponds and even bog pools. **Reproduction** May–June, in shallow water. About 1,000 eggs in small (5–6 cm, 2 years old) spawning females. **Food** invertebrates and (when larger) small fish. **Value** of no commercial or angling value. **Conservation Status** Alien.

Flatfish Family
Pleuronectidae

The Pleuronectidae, or flatfish, is a large family of mainly marine fish found in all seas of the world. There are many genera, but only 2 of these regularly enter fresh water in Europe. As their common name implies, these fish are characteristically flattened and asymmetrical, with both eyes on the same side (usually the right, but reversed specimens occasionally occur).

The development of flatfish is unique. When they hatch from the egg, the larvae look just like those of other fish – bilaterally symmetrical with an eye on each side of the head. However, during late larval development, one eye moves across the head so that both eyes are eventually on the same side of the head. At the same time, the whole shape of the head changes and the fish settles on its side on the sea bed, gradually assuming the adult form.

Only 2 species (both catadromous) occur in fresh water in Europe.

Key to Pleuronectidae in Europe

1	Bony tubercles present at bases of dorsal and anal fin rays; bony tubercles or platelets on body; 15–22 gill rakers on 1st arch	*Platichthys flesus* p249
	No bony tubercles at bases of dorsal and anal fin rays, nor bony tubercles or platelets on body; 10–14 gill rakers on 1st arch	*Pleuronectes glacialis* p249

Atlantic Flounder *Platichthys flesus*

Size 20–30 cm; maximum 50 cm; British rod record 2.594 kg (1956). **Distinctive features** flattened asymmetrically with both eyes on the same side; bony tubercles and platelets on body; 15–22 gill rakers on 1st arch. **Distribution** found all round the European coast from the Arctic Ocean to the northern Mediterranean and Black Seas. Common also in estuaries and lowland rivers and some lakes which are easily accessible from the sea. **Reproduction** February–May, spawning in the sea in deep water. The eggs hatch in 4–8 days and the larvae are pelagic for about 50 days before sinking to the bottom and developing their flattened form. The young mature after 3–4 years. 500,000–2,000,000 eggs per female. **Food** zooplankton when young, benthic invertebrates, especially worms, molluscs and crustaceans when older. **Value** of considerable commercial value in many sandy coastal areas where it is caught in traps and nets of various types. It is also popular with anglers in many coastal areas. **Conservation Status** Lower Risk.

Arctic Flounder *Pleuronectes glacialis*

Size 15–25 cm; maximum 35 cm. **Distinctive features** flattened asymmetrically with both eyes on the same side; no bony tubercles or platelets on body; 10–14 gill rakers on 1st arch. **Distribution** found only in Arctic coastal waters including estuaries and the lower reaches of some rivers. **Reproduction** February–May, in the sea. **Food** benthic invertebrates, especially molluscs. **Value** of local commercial value to coastal net fisheries. Of no sporting value. **Conservation Status** Lower Risk.

Bibliography

Albuquerque, R. M. 1956. Peixes de Portugal e ilhas adjacentes. *Port. Acta biol.*, B. 5, 1–1164.

Almaca, C. 1965. Contribution a la connaissance des poissons des eaux interieures du Portugal. *Sep. Rev. Fac. Cienc. Lisboa.* 13, 225–262.

Almaca, C. Contemporary changes in Portugese freshwater fish fauna and conservation of autochthonous Cyprinidae. *Roczniki Nauk Rolniczych.* 100, 9–15.

Audobon Society. 1988. *Field guide to North American fishes, whales and dolphins.* Chanticleer Press, New York.

Banarescu, P., Blanc, M., Gaudet, J. L. & Hureau, J. C. 1971. *European inland water fish: a multilingual catalogue.* Fishing News, London.

Bauch, G. 1953. *Die einheimischen Susswasserfische.* Neumann, Berlin.

Berg, L. 1947. *Classification of fish both recent and fossil.* Edwards, Ann Arbor.

Berg, L. S. 1962. *Freshwater fishes of the USSR and adjacent countries.* Israel Program of Scientific Translation, Jerusalem.

Bini, G. 1962. *I pesci delle aque interne d'Italia.* Garzanti, Rome.

Bless, R. 1978. Bestandsanderungen der Fischfauna in der Bundesrepublik Deutschland. *Naturschutz aktuell.* 2, 1–66.

Bone, Q., Marshall, N.B. & Blaxter, J.H.S. 1995. *Biology of fish.* Blackie, Glasgow.

Brandt, A.V. 1975. *Fish catching methods of the world.* Fishing News Books, Farnham.

Bruylants, B., Vandelannoote, A. & Verheyen, R. 1989. *De vissenvan onze Vlaamse beken en rivieren: hun ecologie, verspreidung en bescherming.* Antwerpen, WEL.

Cabo, F. L. 1964. *Los peces de las aguas continentales espanolas.* SNPFC, Madrid.

Cazimier, W.G. 1984. Zeldame vissen, kreeften en krabben in de binnenwateren. *Visserij.* 37, 117–132.

Costa Pereira, N.Da 1995. The freshwater fishes of the Iberian peninsula. *Arq. Mus. Bocage, N.S.* 2, 473–538.

Crivelli, A.J. 1996. *The freshwater fish endemic to the northern Mediterranean region.* Tour du Valat, Arles.

Crivelli, A.J. & Maitland, P.S. (Eds) 1995. Endemic freshwater fishes of the northern Mediterranean region. *Biological Conservation.* 72, 121–337.

Curry-Lindahl, K. 1985. *Vara fiskar. Havs-och sotvattensfiskar i Norden och ovriga Europa.* Norstedts.

Delmastro, G. 1982. *I pesci del bacino del Po.* Milano, CLESAV.

Doadrio, I., Elvira, B. & Bernat, Y. (Eds) 1991. *Peces continentales espanoles. Inventario y clasificacion de zonas fluviales.* Madrid, ICONA.

Dottrens, E. 1952. *Les poissons d'eau douce.* Delachaux & Niestle, Neuchatel.

Economidis, P.S. & Nalbant, T.T. 1997. A study of the loaches of the genera *Cobitis* and *Sabanejewia* (Pisces, Cobitidae) of Greece, with description of six new taxa. *Trav. Mus. Natl. Hist. nat. 'Grigore Antipa'.* 36, 295–347.

FAO. 1981. *Conservation of the genetic resources of fish: problems and recommendations.* Food & Agricultural Organisation of the United Nations, Rome.

Forneris, G., Paradisi, S. & Specchi, M. 1990. *Pesci d'acqua dolce.* Lorenzini, Udine.

Gandolfi, G., Zerunian, S., Torricelli, P. & Marconato, A. 1990. I pesci delle acque interne italiane. *Ist. Poligrafico e Zecca dello Stato.* 1–617.

Geldiay, R. & Balik, S. 1996. *Turkiye Tatlisu Baliklari.* Ege Universitesi Basimevi, Izmir.

Greenwood, P. H., Miles, R. S. & Patterson, C. 1973. *Interrelationships of fish.* Academic Press, New York.

Hoar, W. S. & Randall, D. J. 1971. *Fish physiology.* Academic Press, New York.

Hoestlandt, H. (Ed) 1991. *The freshwater fishes of Europe. Clupeidae, Anguillidae.* AULA-Verlag, Wiesbaden.

Holcik, J. (Ed) 1986. *The freshwater fishes of Europe. Petromyzontiformes.* AULA-Verlag, Wiesbaden.

Holcik, J. (Ed) 1989. *The freshwater fishes of Europe. Acipenseriformes.* AULA-Verlag, Wiesbaden.

Holcik, J. & Mihalik, J. 1968. *Freshwater fish.* Svoboda, Prague.

Huet, M. 1971. *Textbook of fish culture: breeding and cultivation of fish.* Fishing News, London.

Hutchinson, G. E. 1975. *A treatise on limnology.* Wiley, New York.

Jenkins, R.E. & Burkhead, N.M. 1994. *Freshwater fishes of Virginia.* Bethesda, American Fisheries Society.

Karoly, P. 1989. *Magyarorszag Halai. Biologiajuk es Hasznositasuk.* Akademiai Kiado, Budapest.

Koli, L. 1994. *Suomen Kalaopas.* WSOY, Helsinki.

Kottelat, M. 1997. European freshwater fishes – an heuristic checklist of the freshwater fishes of Europe (exclusive of former USSR), with an introduction for non-systematists and comments on nomenclature and conservation. *Biologia.* 52 (Supplement 5), 1–271.

Ladiges, W. & Vogt, D. 1979. *Die Susswasserfische Europas bis zum Ural und Kaspischen Meer.* Parey, Hamburg.

Lagler, K. F., Bardach, J. E. & Miller, R. R. 1967. *Ichthyology.* Wiley, New York.

Lelek, A. 1980. *Threatened freshwater fish of Europe.* Strasbourg, Council of Europe.

Lever, C. 1997. *Naturalized fishes of the world.* Academic Press, London.

Maitland, P. S. 1972. A key to the freshwater fish of the British Isles with notes on their distribution and ecology. *Sci. Publ. Freshw. Biol. Ass.*, 27, 1–139.

Maitland, P.S. & Campbell, R.N. 1992. *Freshwater fish of the British Isles.* HarperCollins, London.

Maitland, P.S. & Crivelli, A.J. 1996. *Conservation of freshwater fish.* Tour du Valat, Arles.

Marshall, N. B. 1965. *The life of fish.* Weidenfeld & Nicolson, London.

Miller, P.J. & Loates, M.J. 1997. *Fish of Britain and Europe.* HarperCollins, London.

Muus, B. J. & Dahlstrom, P. 1999. *Freshwater fish.* Scandinavian Fishing Year Book, Hedehusene.

Nelson, J.S. 1994. *Fish of the world.* Wiley, New York.

Nielsen, L. 1990. *Fish in colour.* Politikens, Copenhagen.

Nijssen, H. & de Groot, S.J. 1987. *De Vissen van Nederland.* Stichting Uitgeverij van de Koninklijke Nederlandse Natuurhistorische Vereniging, Utrecht.

Nikolsky, G. V. 1963. *The ecology of fish.* Academic Press, London.

Norman, J. R. & Greenwood, P. H. 1975. *A history of fish.* Benn, London.

Phillipart, J.C. & Vranken, M. 1983. *Protegeons nos Poissons.* Duculot, Gembloux.

Phillips, R. & Rix, M. 1985. *Freshwater fish of Britain, Ireland and Europe.* Pan Books, London.

Pivnilman, C. J. 1961. *Poissons d'eau douce.* Lechevalier, Paris.

Spindler, T. 1995. *Fischfauna in Osterreich.* Bundesministerium fur Umweltbundesamt, Wien.

Steinmann, H. 1948. *Die Fische der Scheiz.* Aarau, Saurlaender.

Sterba, G. 1973. *Freshwater fish of the world.* Studio Vista, London.

Van Duijn, C. 1967. *Diseases of fish.* Iliffe, London.

Welch, P. S. 1951. *Limnology.* McGraw-Hill, New York.

Wootton, R.J. 1990. *Ecology of teleost fish.* Chapman Hall, London.

Glossary

adipose fin modified rayless posterior dorsal fin found in the Salmonidae and a few other families (see page 177).

alevin recently hatched stage of a salmonid fish when the yolk sac still protrudes externally.

anadromous maturing in salt water but migrating into fresh water to spawn.

axillary process pointed structure growing from the base of a fin.

benthic bottom living.

bifid forked.

biota flora and fauna of an area.

branchial of the gills.

branchiostegal of the gill covers.

buccal of the mouth or cheek.

catadromous maturing in fresh water but migrating into salt water to spawn.

caudal peduncle base of tail.

chromatophore pigment cell which can be altered in shape to produce colour change (see fig. 28).

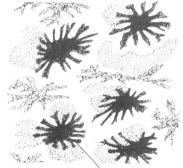

Fig. 28   chromatophore

circuli ring-like ridges on fish scales (see fig. 5c).

ctenoid having a comb-like margin (see fig. 5c).

cusp sharp point or prominence.

cycloid having an evenly curved free border (see page 177).

denticle small tooth-like process.

dentiform tooth-like.

diatom unicellular type of alga with silica walls.

dichotomous repeatedly forking or dividing.

dorso-ventral stretching from dorsal to ventral surface.

emarginate having a notched margin.

epithelial of a covering or enveloping tissue.

erythrocyte red blood corpuscle.

eutrophication increasing in chemical richness.

fimbria delicate fringing processes as found on the barbels of sturgeons.

freshet sudden but temporary increase in flow down a river, often due to heavy rain.

fusiform tapering gradually at both ends.

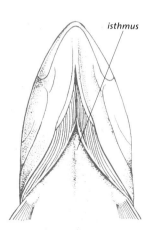

Fig. 29

gene the unit of heredity found in living cells.

gene pool total hereditary material available within a population of animals or plants.

heterocercal tail fin in which the upper lobe is larger than the lower, and contains the upturned termination of the vertebral column (see page 72).

holocercal tail fin in which the upper and lower lobes are the same size (see page 18).

hyoid bone lying at the base of the tongue in fishes.

inferior below.

infra-oral below the mouth.

interorbital between the eye sockets.

isthmus narrow piece of tissue connecting two larger structures (see fig. 29).

keeled having a single ventral ridge.

labial referring to structures which are part of the lips or mouth.

lamella thin plate-like structure.

lateral situated on the side.

lateral band strip of colour along the side of the body.

leucocyte colourless blood corpuscle.

lingual of the tongue.

littoral zone of shallow water around the edges of lakes and seas.

lobule small lobe or projection of an organ.

lymphocyte small colourless blood corpuscle.

mandibular of the lower jaw.

maxillary of the upper jaw (see fig. 33).

medio-lateral along the middle of a side.

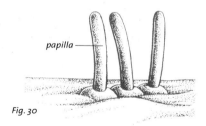

papilla

Fig. 30

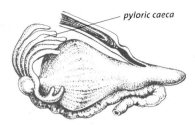

pyloric caeca

Fig. 31

occipital of the back part of the head.
operculum gill cover (see fig. 8).
otolith calcareous particle found in the inner ears of fishes (see fig. 10).
palatine occurring in the region of the palate (see fig. 33).
papilla small projection from the body surface (see fig. 30).
parietal in fishes, usually referring to paired bones on the roof of the skull (see fig. 33).
parr immature stage of a salmonid fish between fry and smolt when a row of distinct dark marks are present along each side.
pelagic zone zone of open water away from the edges and bottom of a lake or sea.
peristalsis movement by means of successive waves of muscular contraction.
pharyngeal of the gullet, or anterior part of the alimentary canal.
photosynthesis the process in green plants where, under the action of light, carbohydrates are synthesized from carbon dioxide and water.
phytoplankton microscopic plants (mainly algae) which drift free in the water.
platelet small flattened disc.
postorbital behind the eye sockets.
preopercular anterior to the gill cover.
preorbital in front of the eye sockets.
pyloric caeca small blind-ending pouches opening to the posterior part of the stomach (see fig. 31).
redd depression in gravel dug out by a

female salmonid fish in which to lay her eggs.
scute bony scale-like structure (see page 72).
smolt immature stage of a salmonid fish following the parr stage when the whole fish becomes completely silver in colour.
suborbital below the eye sockets.
subterminal situated almost, but not quite, at the end of a structure.
superior above.
supra-oral above the mouth.
taxa definite units in the classification of plants and animals.
taxonomist a scientist involved in the classification of plants and animals.
terminal situated at the end of a structure.
tricuspid having three tapering points.
trifid divided to form three lobes.
triserial arranged in three rows.
truncated terminating abruptly.
tubercle small rounded swelling (see fig. 32).

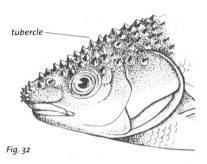

tubercle

Fig. 32

tubule any small hollow cylindrical organ.
turbid cloudy with suspended matter.
unicuspid having one tapering point.
uniserial arranged in one row.
vermiculate marked with numerous bending lines of colour (see *Salvelinus fontinalis*, page 188).
vomer bone in the nasal region (see fig. 3).
zooplankton microscopic animals which drift free in the water.

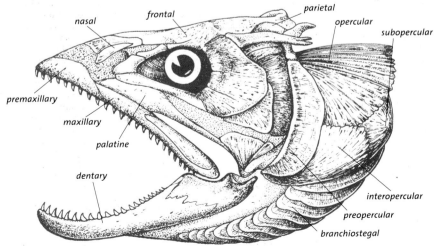

Fig. 33 Principal external bones in the head of a fish (Salmo)

Index

Page numbers in *italic*
refer to illustrations